DYNAMIC MECHANICAL ANALYSIS
A Practical Introduction

Kevin P. Menard

CRC Press

Boca Raton   London   New York   Washington, D.C.

Library of Congress Cataloging-in-Publication Data

Menard, Kevin Peter.
    Dynamic mechanical analysis : a practical introduction / by Kevin
P. Menard.
        p. cm.
    Includes bibliographical references.
    ISBN 0-8493-8688-8 (alk. paper)
    1. Polymers--Mechanical properties. 2. Polymers--Thermal
properties. I. Title.
TA455.P58M45 1999
620.1′9292—dc21
                                                      98-53025
                                                           CIP

No claim to original U.S. Government works
International Standard Book Number 0-8493-8688-8
Library of Congress Card Number 98-53025
Printed in the United States of America 1 2 3 4 5 6 7 8 9 0
Printed on acid-free paper

Preface

As an educator, and also because of my involvement in Short Courses preceding the International Conferences on Materials Characterization (POLYCHAR), I have found repeatedly that some practitioners of polymer science and engineering tend to stay away from dynamic mechanical analysis (DMA). Possibly because of its use of complex and imaginary numbers, such people call the basic DMA definitions impractical and sometimes do not even look at the data. This is a pity, because DMA results are quite useful for the manufacturing of polymeric materials and components as well as for the development of new materials.

Year after year, listening to Kevin Menard's lectures at the International Conference on Polymer Characterization (POLYCHAR) Short Courses on Materials Characterization, I have found that he has a talent for presentation of ostensibly complex matters in a simple way. He is not afraid of going to a toy store to buy slinkies or silly putty — and he uses these playthings to explain what DMA is about. Those lectures and the DMA course he teaches for Perkin-Elmer, which is also part of the graduate-level thermal analysis course he teaches at University of North Texas, form the basis of this text.

The following book has the same approach: explaining the information that DMA provides in a practical way. I am sure it will be useful for both beginning and advanced practitioners. I also hope it will induce some DMA users to read more difficult publications in this field, many of which are given in the references.

Witold Brostow
University of North Texas
Denton, in July 1998

Author's Preface

In the last 5 to 10 years, dynamic mechanical analysis or spectroscopy has left the domain of the rheologist and has becoming a common tool in the analytical laboratory. As personal computers become more and more powerful, this technique and its data manipulations are becoming more accessible to the nonspecialist. However, information on the use of DMA is still scattered among a range of books and articles, many of which are rather formidable looking. It is still common to hear the question "what is DMA and what will it tell me?" This is often expressed as "I think I could use a DMA, but can't justify its cost." Novices in the field have to dig through thermal analysis, rheology, and material science texts for the basics. Then they have to find articles on the specific application. Having once been in that situation, and as I am now helping others in similar straits, I believe there is a need for an introductory book on dynamic mechanical analysis.

This book attempts to give the chemist, engineer, or material scientist a starting point to understand where and how dynamic mechanical analysis can be applied, how it works (without burying the reader in calculations), and what the advantages and limits of the technique are. There are some excellent books for someone with familiarity with the concepts of stress, strain, rheology, and mechanics, and I freely reference them throughout the text. In many ways, DMA is the most accessible and usable rheological test available to the laboratory. Often its results give clear insights into material behavior. However, DMA data is most useful when supported by other thermal data, and the use of DMA data to complement thermal analysis is often neglected. I have tried to emphasize this complementary approach to get the most information for the cost in this book, as budget constraints seem to tighten each year. DMA can be a very cost-effective tool when done properly, as it tells you quite a bit about material behavior quickly.

The approach taken in this book is the same I use in the DMA training course taught for Perkin-Elmer and as part of the University of North Texas course in Thermal Analysis. After a review of the topic, we start off with a discussion of the basic rheological concepts and the techniques used experimentally that depend on them. Because I work mainly with solids, we start with stress–strain. I could as easily start with flow and viscosity. Along the way, we will look at what experimental considerations are important, and how data quality is assured. Data handling will be discussed, along with the risks and advantages of some of the more common methods. Applications to various systems will be reviewed and both experimental concerns and references supplied.

The mathematics has been minimized, and a junior or senior undergraduate or new graduate student should have no trouble with it. I probably should apologize now to some of my mentors and the members of the Society of Rheology for what may be oversimplifications. However, my experience suggests that most users of

DMA don't want, may not need, and are discouraged by an unnecessarily rigorous approach. For those who do, references to more advanced texts are provided. I do assume some exposure to thermal analysis and a little more to polymer science. While the important areas are reviewed, the reader is referred to a basic polymer text for details.

Kevin P. Menard
U. North Texas
Denton, Texas

About the Author

Kevin P. Menard is a chemist with research interests in materials science and polymer properties. He has published over 50 papers and/or patents. Currently a Senior Product Specialist in Thermal Analysis for the Perkin-Elmer Corporation, he is also an Adjunct Professor in Materials Science at the University of North Texas. After earning his doctorate from the Wesleyan University and spending 2 years at Rensselaer Polytechnic Institute, he joined the Fina Oil and Chemical Company. After several years of work on toughened poly-  mers, he moved to the General Dynamics Corporation, where he managed the Process Engineering Group and Process Control Laboratories. He joined Perkin-Elmer in 1992.

Dr. Menard is a Fellow of the Royal Society of Chemistry and a Fellow of the American Institute of Chemists. He is active in the Society of Plastic Engineers, where he is a member of the Polymer Analysis Division Board of Directors. He has been treasurer for the North American Thermal Analysis Society, a local officer of the American Chemical Society, and is a Certified Professional Chemist.

Acknowledgments

I need to thank and acknowledge the help and support of a lot of people, more than could be listed here. This book would never have been started without Dr. Jose Sosa. After roasting me extensively during my job interview at Fina, Jose introduced me to physical polymer science and rheology, putting me through the equivalent of a second Ph.D. program while I worked for him. One of the best teachers and finest scientists I have met, I am honored to also consider him a friend. Dr. Letton and Dr. Darby at Texas A&M got me started in their short courses. Jim Carroll and Randy O'Neal were kind enough to allow me to pursue my interests in DMA at General Dynamics, paying for classes and looking the other way when I spent more time running samples than managing that lab. Charles Rohn gave me just tons of literature when I was starting my library. Chris Macosko's short course and its follow-up opened the mathematical part of rheology to me.

Witold Brostow of the University of North Texas, who was kind enough to preface and review this manuscript, has been extremely tolerant of my cries for help and advice over the years. While he runs my tail off with his International Conference on Polymer Characterization each winter, his friendship and encouragement (translation: nagging) was instrumental in getting this done. Dr. Charles Earnest of Berry College has also been more than generous with his help and advice. His example and advice in how to teach has been a great help in approaching this topic.

My colleagues at the Perkin-Elmer Corporation have been wonderfully supportive. Without my management's support, I could have never done this. John Dwan and Eric Printz were supportive and tolerant of the strains in my personality. They also let me steal shamelessly from our DMA training course I developed for PE. Dr. Jesse Hall, my friend and mentor, has supplied lots of good advice. The TEA Product Department, especially Sharon Goodkowsky, Lin Li, Greg Curran, and Ben Twombly, was extremely helpful with data, advice, samples, and support. Sharon was always ready with help and advice. My counterparts, Dave Norman and Farrell Summers, helped with examples, juicy problems, and feedback. A special thanks goes to the salesmen I worked with: Drew Davis, Peter Muller, Jim Durrett, Ray Thompson, Steve Page, Haidi Mohebbi, Tim Cuff, Dennis Schaff, and John Minnucci, who found me neat examples and interesting problems. Drew deserves a special vote of thanks for putting up with me in what he still believes is his lab. Likewise, our customers, who are too numerous to list here, were extremely generous with their samples and data. I thank Dr. John Enns for his efforts in keeping me honest over the years and his pushing the limits of the current commercially available instrumentation. John Rose of Rose Consulting has been always a source of interesting problems and wide experience. In addition, he proofread the entire manuscript for me. Nandika D'Sousa of UNT also reviewed a draft copy and made helpful suggestions. A very special thanks goes to Professor George Martin of Syracuse

University. Dr. Martin was kind enough to proofread and comment extensively on the initial draft, and many of his suggestions were used. I feel this book was greatly improved by incorporating their comments, and they have my heartfelt thanks. Many deserving people cannot be mentioned, as I promised not to tell where the samples came from.

More personally, Professor Paul R. Buitron III and Dr. Glenn Morris were constant sources of encouragement and practical advice. Paul especially was a great example, and it is largely due to him that I stayed vaguely sane during this effort. Matthew MacKay, John Essa, and Tom Morrissey also helped with their good advice and support. Felicia Shapiro, my editor, put up endlessly with my lack of a concept of deadline. Finally, thanks are offered to my wife, Connie, and my sons, Noah and Benjamin, for letting me write this on nights when I should have been being an attentive husband and father. I promise to stop spending all my time on the computer now so the boys can have their turn.

Dedication

Слáва ôц҃ꙋ н̂ сн҃ꙋ н̂ ст҃омꙋ д҃х҃ꙋ н̂ нꙑн҇ѣ
н̂ прн́снш н̂ во в҇ѣ́ки в҇ѣкш҃вz

To my wife, Connie,
Tecum vivere amen,
tecum obeam libens.
Homer, Epodes, ix

And to Dr. Jose Sosa,
My teacher, mentor, and friend.

Table of Contents

1 An Introduction to Dynamic Mechanical Analysis

Dynamic mechanical analysis (DMA) is becoming more and more commonly seen in the analytical laboratory as a tool rather than a research curiosity. This technique is still treated with reluctance and unease, probably due to its importation from the field of rheology. Rheology, the study of the deformation and flow of materials, has a reputation of requiring a fair degree of mathematical sophistication. Although many rheologists may disagree with this assessment,[1] most chemists have neither the time nor the inclination to delve through enough literature to become fluent. Neither do they have an interest in developing the constituent equations that are a large part of the literature. However, DMA is a technique that does not require a lot of specialized training to use for material characterization. It supplies information about major transitions as well as secondary and tertiary transitions not readily identifiable by other methods. It also allows characterization of bulk properties directly affecting material performance.

Depending on whom you talk to, the same technique may be called dynamic mechanical analysis (DMA), forced oscillatory measurements, dynamic mechanical thermal analysis (DMTA), dynamic thermomechanical analysis, and even dynamic rheology. This is a function of the development of early instruments by different specialties (engineering, chemistry, polymer physics) and for different markets. In addition, the names of early manufacturers are often used to refer to the technique, the same way that "Kleenex™" has come to mean "tissues." In this book, DMA will be used to describe the technique of applying an oscillatory or pulsing force to a sample.

1.1 A BRIEF HISTORY OF DMA

The first attempts to do oscillatory experiments to measure the elasticity of a material that I found was by Poynting in 1909.[2] Other early work gave methods to apply oscillatory deformations by various means to study metals[3] and many early experimental techniques were reviewed by the Nijenhuis in 1978.[4] Miller's book on polymer properties[5] referred to dynamic measurements in this early discussion of molecular structure and stiffness. Early commercial instruments included the Weissenberg Rheogoniometer (~1950) and the Rheovibron (~1958). The Weissenberg Rheogoniometer, which dominated cone-and-plate measurements for over 20 years following 1955, was the commercial version of the first instrument to measure normal forces.[6] By the time Ferry wrote *Viscoelastic Properties of Polymers* in 1961,[7] dynamic measurements were an integral part of polymer science, and he gives the best development of the theory available. In 1967, McCrum et al. collected the current information on DMA and DEA (dielectric analysis) into their landmark textbook.[8] The technique remained fairly specialized until the late 1960s, when commercial instruments became more user-friendly. About 1966, J. Gilham developed the Torsional Braid Analyzer[9] and started the modern period of DMA. In 1971, J. Starita and C. Macosko[10] built a DMA that measured normal forces,[10] and from this came the Rheometrics Corporation. In 1976, Bohlin also develop a commercial DMA and started Bohlin Rheologia. Both instruments used torsional geometry. The early instruments were, regardless of manufacturer, difficult to use, slow, and limited in their ability to process data. In the late 1970s, Murayama[11] and Read and Brown[12] wrote books on the uses of DMA for material characterization. Several thermal and rheological companies introduced DMAs in the same time period, and currently most thermal and rheological vendors offer some type of DMA. Polymer Labs offered a dynamic mechanical thermal analyzer (DMTA) using an axial geometry in the early 1980s. This was soon followed an instrument from Du Pont. Perkin-Elmer developed a controlled stress analyzer based on their thermomechanical analyzer (TMA) technology, which was designed for increased low-end sensitivity. The competition between vendors has led to easier to use, faster, and less expensive instruments. The revolution in computer technology, which has so affected the laboratory, changed the latter, and DMA of all types became more user-friendly as computers and software evolved. We will look at instrumentation briefly in Chapter 4.

1.2   BASIC PRINCIPLES

DMA can be simply described as *applying an oscillating force to a sample and analyzing the material's response to that force* (Figure 1.1). This is a simplification, and we will discuss it in Chapter 4 in greater detail. From this, one calculates properties like the tendency to flow (called viscosity) from the phase lag and the stiffness (modulus) from the sample recovery. These properties are often described as the ability to lose energy as heat (damping) and the ability to recover from deformation (elasticity). One way to describe what we are studying is the relaxation of the polymer chains.[13] Another way would be to discuss the changes in the free volume of the polymer that occur.[14] Both descriptions allow one to visualize and describe the changes in the sample. We will discuss stress, strain, and viscosity in Chapter 2.

The applied force is called stress and is denoted by the Greek letter, σ. When subjected to a stress, a material will exhibit a deformation or strain, γ. Most of us working with materials are used to seeing stress–strain curves as shown in Figure 1.2. These data have traditionally been obtained from mechanical tensile testing at a fixed temperature. The slope of the line gives the relationship of stress to strain and is a measure of the material's stiffness, the modulus. The modulus is dependent on the temperature and the applied stress. The modulus indicates how well a material will work in specific application in the real world. For example, if a polymer is heated so that it passes through its glass transition and changes from glassy to rubbery, the modulus will often drop several decades (a decade is

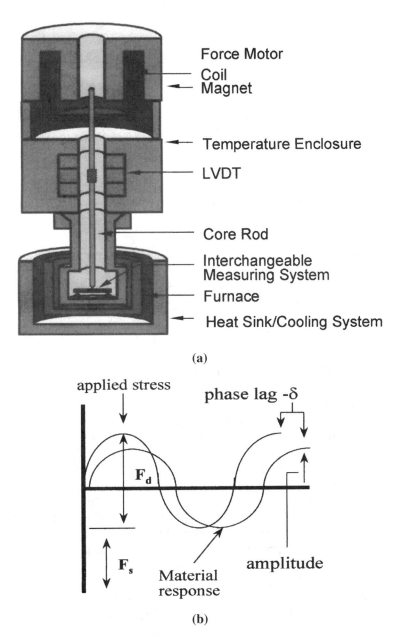

FIGURE 1.1 How a DMA works. The DMA supplies an oscillatory force, causing a sinusoidal stress to be applied to the sample, which generates a sinusoidal strain. By measuring both the amplitude of the deformation at the peak of the sine wave and the lag between the stress and strain sine waves, quantities like the modulus, the viscosity, and the damping can be calculated. The schematic above shows the Perkin-Elmer DMA 7e: other instruments use force balance transducers and optical encoders to track force or position. F_d is the dynamic or oscillatory force while F_s is the static or clamping force. (Used with the permission of the Perkin-Elmer Corporation, Norwalk, CT.)

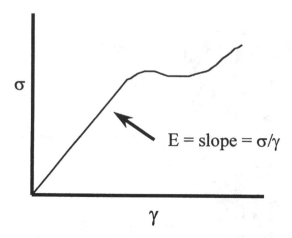

FIGURE 1.2 Stress–strain curves relate force to deformation. The ratio of stress to strain is the modulus (E), which is a measurement of the material's stiffness, or its resistance to deformation. Young's modulus, the slope of the initial linear portion of the stress–strain curve, is commonly used as indicator of material performance in many industries. Since stress–strain experiments are one of the simplest tests for stiffness, Young's modulus provides a useful evaluation of material performance.

an order of magnitude). This drop in stiffness can lead to serious problems if it occurs at a temperature different from expected. One advantage of DMA is that we can obtain a modulus each time a sine wave is applied, allowing us to sweep across a temperature or frequency range. So if we were to run an experiment at 1 Hz or 1 cycle/second, we would be able to record a modulus value every second. This can be done while varying temperature at some rate, such as 10°C/min, so that the temperature change per cycle is not significant. We can then with a DMA record the modulus as a function of temperature over a 200°C range in 20 minutes. Similarly, we can scan a wide frequency or shear rate range of 0.01 to 100 Hz in less than 2 hours. In the traditional approach, we would have to run the experiment at each temperature or strain rate to get the same data. For mapping modulus or viscosity as a function of temperature, this would require heating the sample to a temperature, equilibrating, performing the experiment, loading a new sample, and repeating at a new temperature. To collect the same 200°C range this way would require several days of work.

The modulus measured in DMA is, however, not exactly the same as the Young's modulus of the classic stress–strain curve (Figure 1.3). Young's modulus is the slope of a stress–strain curve in the initial linear region. In DMA, a complex modulus (E^*), an elastic modulus (E'), and an imaginary (loss) modulus (E'')[15] are calculated from the material response to the sine wave. These different moduli allow better characterization of the material, because we can now examine the ability of the material to return or store energy (E'), to its ability to lose energy (E''), and the ratio of these effects (tan delta), which is called damping. Chapter 4 discusses dynamic moduli along with how DMA works.

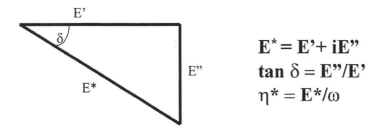

$$E^* = E' + iE''$$
$$\tan \delta = E''/E'$$
$$\eta^* = E^*/\omega$$

FIGURE 1.3 DMA relationships. DMA uses the measured phase angle and amplitude of the signal to calculate a damping constant, D, and a spring constant, K. From these values, the storage and loss moduli are calculated. As the material becomes elastic, the phase angle, δ, becomes smaller, and E^* approaches E'.

Materials also exhibit some sort of flow behavior, even materials we think of as solid and rigid. For example, the silicon elastomer sold as Silly Putty™ will slowly flow on sitting even though it feels solid to the touch. Even materials considered rigid have finite although very large viscosity and "if you wait long enough everything flows[16]." Now to be honest, sometimes the times are so long as to be meaningless to people but the tendency to flow can be calculated. However, this example illustrates that the question in rheology is not if things flow, but how long they take to flow. This tendency to flow is measured as *viscosity*. Viscosity is scaled so it increases with resistance to flow. Because of how the complex viscosity (η^*) is calculated in the DMA, we can get this value for a range of temperatures or frequencies in one scan. The Cox–Mertz rules[17] relate the complex viscosity, η^*, to traditional steady shear viscosity, η_s, for very low shear rates, so that a comparison of the viscosity as measured by dynamic methods (DMA) and constant shear methods (for example, a spinning disk viscometer) is possible.

1.3 SAMPLE APPLICATIONS

Let's quickly look at a couple of examples on using the DMA to investigate material properties. First, if we scan a sample at a constant ramp rate, we can generate a graph of elastic modulus versus temperature. In Figure 1.4a, this is shown for nylon. The glass transition can be seen at ~50° C. Note that there are also changes in the modulus at lower temperatures. These transitions are labeled by counting back from the melting temperature, so the glass transition (T_g) here is also the alpha transition (T_α). As the T_g or T_α can be assigned to gradual chain movement, so can the beta transition (T_β) be assigned to other changes in molecular motions. The beta transition is often associated with side chain or pendant group movements and can often be related to the toughness of a polymer.[18] Figure 1.4b also shows the above nylon overlaid with a sample that fails in use. Note the differences in both the absolute size (the area of the T_β peak in the tan δ) and the size relative to the T_g of T_β. The differences suggest the second material would be much less able to dampen impact via localized chain movements. An idealized scan of various DMA transitions is shown in Figure 1.5,

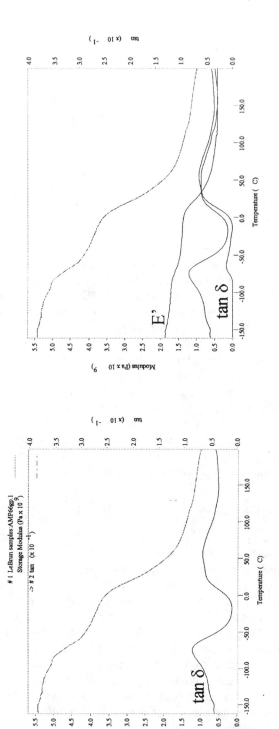

FIGURE 1.4 DMA of a nylon. (a) The importance of higher transitions in material behavior is well known. This sample of material has good impact toughness. We can see in the storage modulus, E', both a T_g at ~50°C and a strong T_β at ~80°. These are also seen as peaks in the tan δ. (b) The curves for the material that fails impact testing are overlaid. Note the lower modulus values and the relatively weaker T_β in the bad sample. Comparisons of the relative peak areas for T_β suggest that the second material is less able to damp vibrations below the T_g.

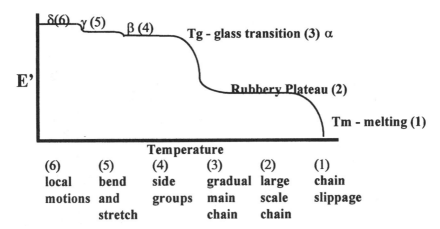

FIGURE 1.5 Idealized DMA scan. An idealized scan showing the effect of various molecular relaxations of the storage modulus, E', curve. In some materials like PET, the beta transition occurs as a broad slope, while in other it exhibits a relativity sharp drop. This is elaborated on in Chapter 5.

along with the molecular motions associated with the transitions. The use of molecular motions and free volume to describe polymer behavior will be discussed in Chapter 5. Another use of this kind of information is determining the operating range of a polymer, for example polyethylene terephthalate (PET). In the range between T_α and T_β, the material possesses the stiffness to resist deformation and the flexibility to not shatter under strain. It is important to note that beta and gamma transitions are too faint to be detected in the differential scanning calorimeter (DSC) or Thermomechanical Analyzer (TMA).[19] The DMA is much more sensitive than these techniques and can easily measure transitions not apparent in other thermal methods. This sensitivity allows the DMA to detect the T_g of highly crosslinked thermosets or of thin coatings.

If we look at a thermoset instead of a thermoplastic, we can follow the material through its cure by tracking either viscosity or modulus changes. This is done for everything from hot melt adhesives to epoxies to angel food cake batter (Figure 1.6). The curves show the same initial decrease in modulus and viscosity to a minimum, corresponding to the initial melting of the uncured material, followed by an increase in viscosity as the material is cured to a solid state. Figure 1.6a shows a cure cycle for an epoxy resin. From one scan, we can estimate the point of gelation (where the material is gelled), the minimum viscosity (how fluid it gets), and when it is stiff enough to bear its own weight.[20] At the last point, we can free up the mold and finish curing in an oven. We can even make a crude relative estimation of the activation energy (E_{act}) from the slope of the viscosity increase during cure.[21] If we want a more exact value for E_{act}, we can use isothermal runs (Figure 1.7) to get values closer to the accuracy of DSC.[22] Chapter 6 looks at these applications in detail.

Often the response of a material to the rate of strain is as important as the temperature response. Chapter 7 addresses the use of frequency scans in the DMA. This is one of the major applications of DMA for polymer melts, suspensions, and solutions. Similarly to how DMA can be used to rapidly map the modulus of a

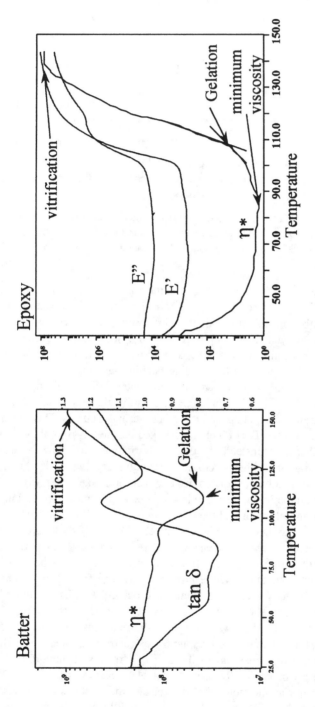

FIGURE 1.6 Curing in the DMA. The curing of very different materials has similar requirements and problems. Note the similarities between a cake batter and an epoxy adhesive. Both show the same type of curing behavior, an initial decrease in viscosity to a minimum followed by a sharp rise to a plateau. Note that gelation is often taken as the E'–E'' crossover or where tan δ = 1. Other points of interest are labeled.

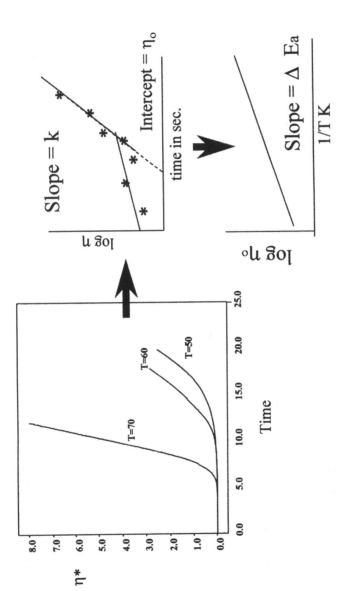

FIGURE 1.7 Isothermal cures for ΔE_a. Isothermal runs allow the development of models for curing. Plotting the log of the measured viscosity, η^*, against time for each temperature gives the true initial viscosity, η_0, and the rate constant, k. Then we obtain the two activation energies, ΔE_a and ΔE_η, by plotting the initial viscosities and rate constants against the inverse temperature ($1/T$). This approach is discussed in Chapter 6.

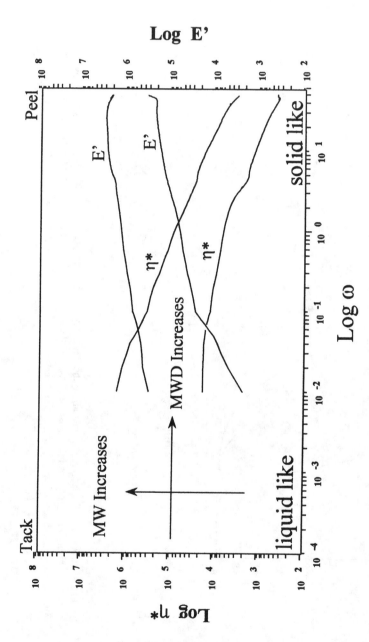

FIGURE 1.8 Frequency scans. Frequency scans are one of the less often used methods in DMA. Frequency responses depend on molecular structure and can be used to probe the molecular weight and distribution of the material. Properties such as relative tack (stickiness) and peel (resistance to removal) responses can also be studied.

material as a function of temperature, we can also use DMA to quickly look at the effect of shear rate or frequency on viscosity. For example, a polymer melt can be scanned in a DMA for the effect of frequency on viscosity in less than 2 hours over a range of 0.01 Hz to 200 Hz. A capillary rheometer study for similar rates would take days. For a hot melt adhesive, we may need to see the low frequency modulus (for stickiness or tack) as well as the high frequency response (for peel resistance).[23] We need to keep the material fluid enough to fill the pores of the substrate without the elasticity getting so low the material pulls out of the pores too easily. By scanning across a range of frequencies (Figure 1.8), we can collect information about the elasticity and flow of the adhesive as E' and η^* at the temperature of interest.

The frequency behavior of materials can also give information on molecular structure. The crossover point between either E' and η^* or between E' and E'' can be related to the molecular weight[24] and the molecular weight distribution[25] by the Doi–Edwards theory. As a qualitative assessment of two or more samples, this crossover point allows a fast comparison of samples that may be difficult or impossible to dissolve in common solvents. In addition, the frequency scan at low frequency will level off to the zero-shear plateau (Figure 1.9). In this region, changes in frequency do not result in a change in viscosity because the rate of deformation is too low for the chains to respond. A similar effect, the infinite shear plateau, is found at very high frequencies. The zero-shear plateau viscosity can be directly related to molecular weight, above a critical molecular weight by

$$\eta = k\left(M^{3.4}\right) \tag{1.1}$$

where k is a material specific constant. This method has been found to be as accurate as gel permeation chromatography (GPC) over a very wide range of molecular weights for the polyolefins.[27]

Frequency data are often manipulated in various ways to extend the range of the analysis by exploiting the Boltzmann superposition principle.[28] Master curves from superpositioning strain, frequency, time, degree of cure, humidity, etc., allow one to estimate behavior outside the range of the instrument or of the experimenter's patience.[29] Like all accelerated aging and predictive techniques, one needs to remember that this is a bit like forecasting the weather, and care is required.[17]

1.4　CREEP–RECOVERY TESTING

Finally, most DMAs on the market also allow creep–recovery testing. Creep is one of the most fundamental tests of material behavior and is directly applicable to a product performance.[30] We discuss this in Chapter 3 as part of the review of basic principles, as it is the basic way to study polymer relaxation. Creep–recovery testing is also a very powerful analytical tool. These experiments allow you to examine a material's response to constant load and its behavior on removal of that load. For example, how a cushion on a chair responds to the body weight of the occupant, how long it takes to recover, and how many times it can be sat on before it becomes permanently compressed can all be studied by creep–recovery testing. The creep

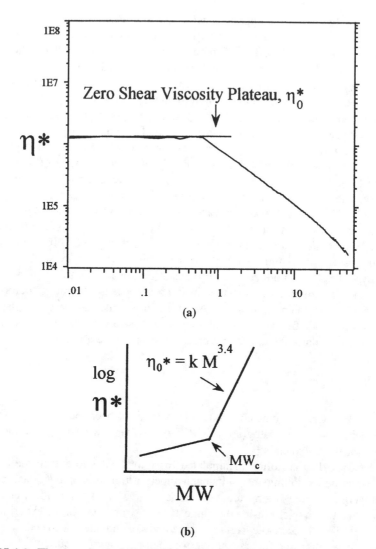

(a)

(b)

FIGURE 1.9 The zero shear plateau. One of the main uses of frequency data is estimation of molecular weight. The zero shear plateau can be used to calculate the molecular weight of a polymer by the above equation if the material constant k is known and the MW is above a critical value. This critical molecular weight, M_c, is typically about 10,000 amu.

experiment can also be used to collect data at very low frequencies[31] and the recovery experiment to get data at high frequencies by free oscillations,[32] extending the range of the instrument. This is discussed in Sections 3.3 and 4.3, respectively. More importantly, creep–recovery testing allows you to gain insight into how a material will respond when kept under constant load, such as a plastic wheel on a caster.

Note that creep is not a dynamic test, as a constant load is applied during the creep step and removed for the recovery step (Figure 1.10). Several approaches to

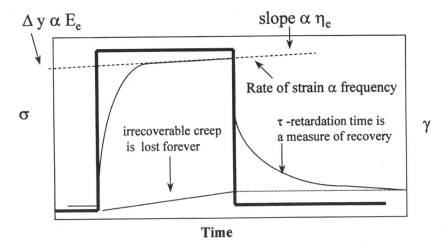

FIGURE 1.10 Creep–recovery testing. Creep–recovery experiments allow the determination of properties at equilibrium like modulus, E_e, and viscosity, η_e. These values allow the prediction of material behavior under conditions that mimic real life applications.

quantifying the data can be used, as shown in Figure 1.10,[33] and will be discussed in Chapter 3. Comparing materials after multiple cycles can be used to magnify the differences between materials as well as predict long-term performance (Figure 1.11). Repeated cycles of creep–recovery show how the product will wear in the real world, and the changes over even three cycles can be dramatic. Other materials, such as a human hair coated with commercial hair spray, may require testing for over a hundred cycles. Temperature programs can be applied to make the test more closely match what the material is actually exposed to in end use. This can also be done to accelerate aging in creep studies by using oxidative or reductive gases, UV exposure, or solvent leaching.[34]

1.5 ODDS AND ENDS

Any of these tests mentioned above can be done in controlled-environment conditions to match the operating environment of the samples. Examples include hydrogels tested in saline,[35] fibers in solutions,[36] and collagen in water.[37] UV light can be used to cure samples[38] to mimic processing or operating conditions. A specialized example of environmental testing is shown in Figure 1.12, where the position control feature of a DMA is exploited to perform a specialized stress relaxation experiment called constant gauge length (CGL) testing. The response of the fibers is greatly affected by the solution it is tested in. Similar tests in both dynamic and static modes are used in the medical, automotive, and cosmetic industries. The adaptability of the DMA to match real-world conditions is yet another advantage of the technique. The DMA's ability to give insight into the molecular structure and to predict in-service performance makes it a necessary part of the modern thermal laboratory.

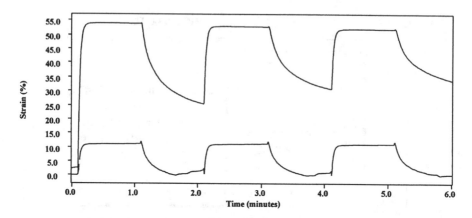

Note the change in samples over three cycles. Relaxation time
increases and percent recovery decreases.

FIGURE 1.11 Creep over multiple cycles. Various programs can be used to simulate the
in-service stressing of a sample, including multiple cycles, temperature changes, and other
environmental factors. Here, the specimen is loaded three times, and the changes in sample
response over three cycles are significant. The relaxation time increases and percent recovery
decreases. This could lead to poor performance if this product is used under repetitive
applications of load.

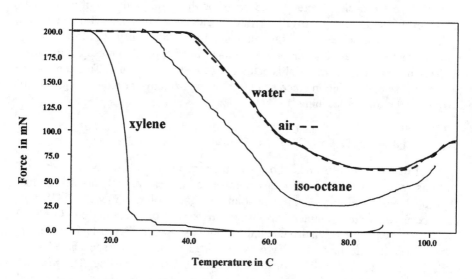

FIGURE 1.12 Environmental conditions affect properties. Testing in the presence of
solvents allows one to evaluate a material under operating conditions. Polypropylene fibers
show very different responses when run in different solvents in a constant gauge length
experiment.

Finally, we will very briefly look at putting all this together by deciding which test to run, how to validate the data we collect, and exploiting other techniques that complement the DMA. Several thermal, spectroscopic, and mechanical tests can be used to help interpret the data. A quick overview of these is given in Chapter 8, along with some guidelines on using DMA tests.

NOTES

1. C. Mascosko, *Rheology Principles, Measurements, and Applications,* VCH, New York, 1994.
2. J. H. Poynting, *Proceedings of the Royal Society, Series A,* 82, 546, 1909.
3. A. Kimball and D. Lovell, *Trans. Amer. Soc. Mech. Eng.,* 48, 479, 1926.
4. K. te Nijenhuis, *Rheology,* Vol. 1, *Principles,* G. Astarita et al., Eds., Plenum Press, New York, 263, 1980.
5. M. L. Miller, *The Structure of Polymers,* Reinhold, New York, 1966.
6. J. Dealy, *Rheometers for Molten Plastics,* Van Nostrand Reinhold, New York, 136–137, 234–236, 1992.
7. J. Ferry, *Viscoelastic Properties of Polymers,* 3rd ed., Wiley, New York, 1980.
8. N. McCrum, B. Williams, and G. Read, *Anelastic and Dielectric Effects in Polymeric Solids,* Dover, New York, 1991. (Reprint of the 1967 edition.)
9. J. Gilham and J. Enns, *Trends in Polymer Science,* 2, 406, 1994.
10. C. Macosko and J. Starita, *SPE Journal,* 27, 38, 1971.
11. T. Murayama, *Dynamic Mechanical Analysis of Polymeric Materials,* Elsevier, New York, 1977. This book is the ultimate reference on the Rheovibron.
12. B. E. Read and G. D. Brown, *The Determination of the Dynamic Properties of Polymers and Composites,* Wiley, New York, 1978.
13. S. Matsuoka, *Relaxation Phenomena in Polymers,* Hanser, New York, 1992.
14. W. Brostow and R. Corneliussen, Eds., *Failure of Plastics,* Hanser, New York, 1986.
15. N. McCrum, B. Williams, and G. Read, *Anelastic and Dielectric Effects in Polymeric Solids,* Dover, New York, 1991.
16. H. Barnes, J. Hutton, and K. Walters, *An Introduction to Rheology,* Elsevier, New York, 1989.
17. J. Dealy and K. Wissbrum, *Melt Rheology and Its Role in Plastics Processing,* Van Nostrand, New York, 1990.
18. This is admittedly a generalization of a very complex subject. B. Twombly, K. Fielder, R. Cassel, and W. Brennan, *NATAS Proceeding,* 20, 28, 1991. D. Van Krevelen, *Properties of Polymers,* Elsevier, New York, 1972. R. Boyd, *Polymer,* 26, 323, 1123, 1985. N. McCrum, B. Williams, and G. Read, *Anelastic and Dielectric Effects in Polymeric Solids,* Dover, New York, 1991.
19. R. Cassel and B. Twombly in *Material Characterization by Thermomechanical Analysis,* M. Neag, Ed., ASTM, Philadelphia, STP 1136, 108, 1991.
20. S. Crane and B. Twombly, *NATAS Proceedings,* 20, 386, 1991.
21. K. Hollands and I. Kalnin, *Adv. Chem. Ser.,* 92, 80, 1970.
22. M. Roller, *Polym. Eng. Sci.,* 15 (6), 406, 1975.
23. C. Rohm, *Proc. of the 1988 Hot Melt Symposium,* 77, 1988.
24. R. Rahalkar and H. Tang, *Rubber Chemistry and Technology,* 61 (5), 812, 1988. W. Tuminello, *Polym. Eng. Sci.,* 26 (19), 1339, 1986.

25. R. Rahalkar, *Rheologica Acta,* 28, 166, 1989. W. Tuminello, *Polym. Eng. Sci.,* 26 (19), 1339, 1986.

26. L. Sperling, *Introduction to Physical Polymer Science Second Edition,* Academic Press, New York, 1993.

27. J. Sosa and J. Bonilla, private communication. B. Shah and R. Darby, *Polym. Eng. Sci.,* 22 (1), 53, 1982.

28. J. Ferry, *Viscoelastic Properties of Polymers,* Wiley, New York, 1980.

29. A. Goldman, *Prediction of the Properties of Polymeric and Composite Materials,* ACS, Washington, 1994. W. Brostow and R. Corneliussen, Eds., *Failure of Plastics,* Hanser, New York, 1986.

30. L. Nielsen, *Mechanical Properties of Polymers,* Reinhold, New York, Ch. 4, 1965.

31. L. Nielsen, *Mechanical Properties of Polymers and Composites,* Marcel Dekker, New York, vol. 1, 1974.

32. U. Zolzer and H. Eicke, *Rheologica Acta,* 32, 104, 1993.

33. L. Nielsen, *Mechanical Properties of Polymers,* vol. 2, Reinhold, New York, 1965. L. Nielsen, *Polymer Rheology,* Marcel Dekker, New York, 1977.

34. Y. Goldman, *Predication of Polymer Properties and Performance,* American Chemical Society, Washington, D.C., 1994.

35. Q. Bao, *NATAS Proceedings,* 21, 606, 1992. J. Enns, *NATAS Proceedings,* 23, 606, 1994.

36. C. Daley and K. Menard, *SPE Technical Papers,* 39, 1412, 1994. C. Daley and K. Menard, *NATAS Notes,* 26 (2), 56, 1994.

37. B. Twombly, R. Cassel, and A. Miller, *NATAS Proceedings,* 23, 288, 1994.

38. J. Enns, unpublished results.

2 Basic Rheological Concepts Stress, Strain, and Flow

The term rheology seems to generate a slight sense of terror in the average worker in materials. Rheology is defined as the study of the deformation and flow of materials. The term was coined by Bingham to describe the work being done in modeling how materials behave under heat and force. Bingham felt that chemists would be frightened away by the term "continuum mechanics," which was the name of the branch of physics concerned with these properties.[1] This renaming was one of science's less successful marketing ploys, as most chemists think rheology is something only done by engineers with degrees in non-Newtonian fluid mechanics, and mostly likely in dark rooms.

More seriously, rheology does have an undeserved reputation of requiring a large degree of mathematical sophistication. Parts of it do require a familiarity with mathematics. However, many of the principles and techniques are understandable to anyone who survived physical chemistry. Enough understanding of its principles to use a DMA and successfully interpret the results doesn't require even that much. For those who find they would like to see more of the field, Chris Mascosko of University of Minnesota has published the text of his short course,[2] and this is a very readable introduction. We are now going to take a nonmathematical look at the basic principles of rheology so we have a common terminology to discuss DMA. This discussion is an elaboration of a lecture given by the author at University of Houston called "Rheology for the Mathematically Insecure."

2.1 FORCE, STRESS, AND DEFORMATION

If you apply a force to a sample, you get a deformation of the sample. The force, however, is an inexact way of measuring the cause of the distortion. Now we know the force exerted by a probe equals the mass of the probe (m) times its acceleration (a). This is given as

$$F = m * a \qquad (2.1)$$

However, force doesn't give the true picture, or rather it gives an incomplete picture. For example, which would you rather catch? A 25-lb medicine ball or a hardball with 250 ft-lb of force? The above equation tells the force exerted by a 25-lb medicine ball moving at 10 ft/s is equal to the force exerted by a hardball (0.25 lb) at 1000 ft/s! We expect that the impact from the hardball is going to do more damage. This damage is called a deformation.

So what we really need to include is a measure of the area of impact for the example above. The product of a force (F) across an area (A) is a stress, σ. The stress is normally given in either pounds per square inch (psi) or pascals and can be calculated for the physical under study. If we assume the medicine ball has an area of impact equivalent to a 1-ft diameter, while the superball's is 1 in., we can calculate the stress by

$$\sigma = F * A \tag{2.2}$$

Figure 2.1 gives a visual interpretation of the results. Despite the forces being equal at 250 ft-lb, the stress seen by the sample varies from 250 psi to 2000 psi. If we program a test in forces and our samples are not exactly the same size, the results will differ because the stresses are different. This occurs even in the linear viscoelastic region,[3] though in the nonlinear region the effects are magnified. While stress gives a more realistic view of the material environment, in real life most people choose to operate their instruments in forces. This allows you to track your instruments limits more easily than having to remember the maximum stress in each geometry would. However, it is important to realize that stress is what is affecting the material, not force.

The applied stress causes a deformation of the material, and this deformation is called a strain, γ. A mnemonic for the difference is to remember that if your boss is under a lot of stress, his personality changes are a strain. Strain is calculated as

$$\gamma = \Delta Y / Y \tag{2.3}$$

where Y is the original sample dimension and ΔY is the change in that dimension under stress. This is often multiplied by 100 and expressed as percent of strain. Applying a stress to a sample and recording the resultant strain is a commonly used technique. Going back to the balls above, the different stresses can cause very different strains in the material, be it a sheet of plastic or your catching hand. Figure 2.2 shows how some strains are calculated.

2.2 APPLYING THE STRESS

How the stress is applied can also affect the deformation of the material. Without considering yet how changes in the material's molecular structure or processing enter the picture, we can see different behavior depending how a static stress[4] is applied and what mode of deformation is used. Figure 2.3 summarizes the main ways of applying a static stress. If we consider the basic stress–strain experiment from materials testing, we increase the applied stress over time at a constant stress rate ($\dot{\sigma}$).[5] As shown in Figure 2.3a, we can track the change in strain as a function of stress. This is done at constant temperature and generates the stress–strain curve. Classically, this experiment was done by a strain-controlled instrument, where one

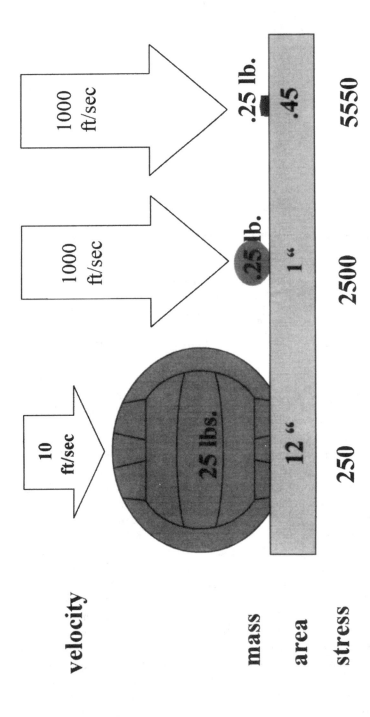

FIGURE 2.1 Force vs. stress. Stress is force divided by area: While the force is a constant at 250 ft-lb, the stress changes greatly as the area changes.

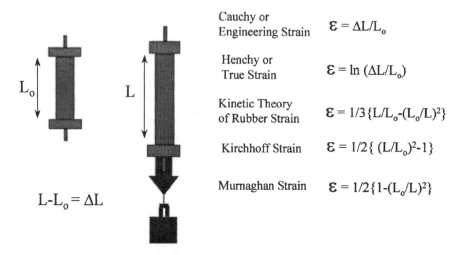

$$\varepsilon = \Delta L / L_o \qquad \text{Cauchy or Engineering Strain}$$

$$\varepsilon = \ln (\Delta L / L_o) \qquad \text{Henchy or True Strain}$$

$$\varepsilon = 1/3 \{ L/L_o - (L_o/L)^2 \} \qquad \text{Kinetic Theory of Rubber Strain}$$

$$\varepsilon = 1/2 \{ (L/L_o)^2 - 1 \} \qquad \text{Kirchhoff Strain}$$

$$\varepsilon = 1/2 \{ 1 - (L_o/L)^2 \} \qquad \text{Murnaghan Strain}$$

$$L - L_o = \Delta L$$

FIGURE 2.2 The result of applying a stress is a strain. There are many types of strain, developed for specific problems.

deforms the sample at a constant strain rate, $\dot{\gamma}$, and measures the stress with a load cell resulting a stress (y-axis)–strain (x-axis) curve. Alternatively, we could apply a constant stress as fast as possible and watch the material deform under that load. This is the classical engineering creep experiment. If we also watch what happens when that stress is removed, we have the creep–recovery experiment (Figure 2.3b). These experiments complement DMA and are discussed in Chapter 3. Conversely, in the stress relaxation experiment, a set strain is applied and the stress decrease over time is measured. Finally, we could apply a constant force or stress and vary the temperature while watching the material change (Figure 2.3c). This is a TMA (thermomechanical analysis) experiment. Thermomechanical analysis is often used to determine the glass transition (T_g) in flexure by heating a sample under a constant load. The heat distortion test used in the polymer industry is a form of this. Figure 2.4 shows the strain response for these types of applied stresses.

While we are discussing applying a stress to generate a strain, you can also look at it as if an applied strain has an associated stress. This approach was often used by older controlled deformation instruments; for example, most mechanical testers used for traditional stress–strain curves and failure testing. The analyzers worked by using a mechanical method, such as a screw drive, to apply a set rate of deformation to a sample and measured the resulting stress with a load cell. The differences between stress control and strain control are still being discussed, and it has been suggested that stress control, because it gives the critical stress, is more useful.[6] We will assume for most of this discussion that the differences are minimal.

The stress can also be applied in different orientations, as shown in Figure 2.5. These different geometries give different-appearing stress–strain curves and, even in an isotropic homogeneous material, give different results. Flexural, compressive, extensile, shear, and bulk moduli are not the same, although some of these can be interconverted, as discussed in Section 2.3 below. For example, in compression a limiting case is reached where the material becomes practically incompressible. This

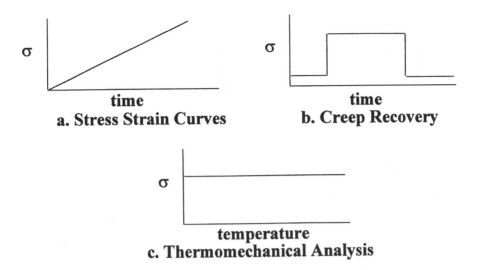

FIGURE 2.3 **Applications of a static stress to a sample.** The three most common cases are shown plotted against time or temperature: (a) stress–strain curves, (b) creep–recovery, (c) thermomechanical analysis.

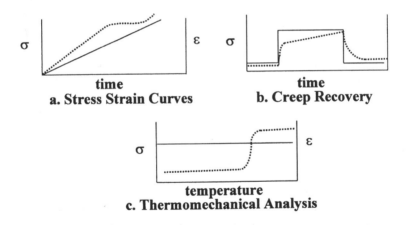

FIGURE 2.4 **Strains resulting from static stress testing.** (a) Stress–strain curves, (b) creep–recovery, (c) thermomechanical analysis. Solid line: stress. Dashed line: strain. For (a), the stress rate is constant. In (c) the probe position rather than strain is normally reported.

limiting modulus is not seen in extension, where the material may neck and deform at high strains. In shear, the development of forces or deformation normal or orthogonal to the applied force may occur.

2.3 HOOKE'S LAW DEFINING THE ELASTIC RESPONSE

Material properties can be conceived of as being between two limiting extremes. The limits of elastic or Hookean behavior[7] and viscous or Newtonian behavior can

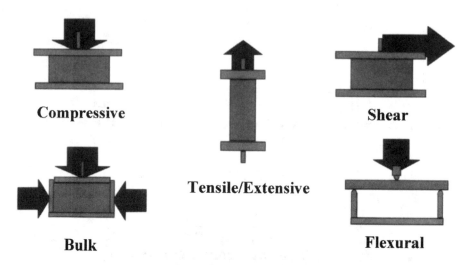

FIGURE 2.5 Modes of deformation. Geometric arrangements or methods of applying stress are shown. All of bottom fixtures are stationary. Bulk is 3-D compression, where sides are restrained from moving.

be looked at as brackets on the region of DMA testing. In traditional stress–strain curves, we are concerned mainly with the elastic response of a material. This behavior can be described as what one sees when you stress a piece of tempered steel to an small degree of strain. The model we use to describe this behavior is the spring, and Hooke's Law relates the stress to the strain of a spring by a constant, k. This is graphically shown in Figure 2.6.

Hooke's law states that the deformation or strain of a spring is linearly related to the force or stress applied by a constant specific to the spring. Mathematically, this becomes

$$\sigma = k * \gamma \tag{2.4}$$

where k is the spring constant. As the spring constant increases, the material becomes stiffer and the slope of the stress–strain curve increases. As the initial slope is also Young's modulus, the modulus would also increase. Modulus, then, is a measure of a material's stiffness and is defined as the ratio of stress to strain. For an extension system, we can then write the modulus, E, as

$$E = d\sigma/d\varepsilon \tag{2.5}$$

If the test is done in shear, the modulus denoted by G and in bulk as B. If we then know the Poisson's ratio, ν, which is a measure of how the material volume changes with deformation when pulled in extension, we can also convert one modulus into another (assuming the material is isotropic) by

$$E = 2G(1 - \nu) = 3B(1 - 2\nu) \tag{2.6}$$

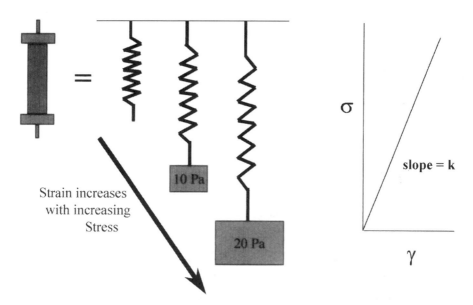

FIGURE 2.6 Hooke's Law and stress–strain curves. Elastic materials show a linear and reversible deformation on applying stress (within the linear region). The slope, k, is the modulus, a measure of stiffness, for the material. For a spring, k would be the spring constant.

For a purely elastic material, the inverse of modulus is the compliance, J. The compliance is a measure of a material's willingness to yield. The relationship of

$$E = 1/J \qquad (2.7)$$

is only true for purely elastic materials, as it does not address viscous or viscoelastic contributions.

Ideally, elastic materials give a linear response where the modulus is independent of load and of loading rate. Unfortunately, as we know, most materials are not ideal. If we look at a polymeric material in extension, we see that the stress–strain curve has some curvature to it. This becomes more pronounced as the stress increases and the material deforms. In extension, the curve assumes a specific shape where the linear region is followed by a nonlinear region (Figure 2.7). This is caused by necking of the specimen and its subsequent drawing out. In some cases, the curvature makes it difficult to determine the Young's modulus.[8]

Figure 2.8 also shows the analysis of a stress–strain curve. Usually, we are concerned with the stiffness of the material, which is obtained as the Young's modulus from the initial slope. In addition, we would like to know how much stress is needed to deform the material.[9] This is the yield point. At some load the material will fail (break), and this is known as the ultimate strength. It should be noted that this failure at the ultimate strength follows massive deformation of the sample. The area under the curve is proportional to the energy needed to break the sample. The shape of this curve and its area tells us about whether the polymer is tough or brittle

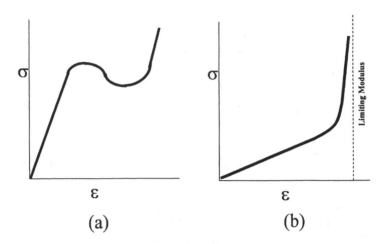

FIGURE 2.7 Stress–strain curves by geometry. The stress–strain curves vary depending on the geometry used for the test. Stress-strain curves for (a) tensile and (b) compressive are shown.

or weak or strong. These combinations are shown in Figure 2.9. Interestingly, as the testing temperature is increased, a polymer's response moves through some or all of these curves (Figure 2.10a). This change in response with temperature leads to the need to map modulus as a function of temperature (Figure 2.10b) and represents another advantage of DMA over isothermal stress–strain curves. Before we discuss that, we need to explain the curvature seen in what Hooke's law says should be a straight line.

2.4   LIQUID-LIKE FLOW OR THE VISCOUS LIMIT

To explain the curvature in the stress–strain curves of polymers, we need to look at the other end of the material behavior continuum. The other limiting extreme is liquid-like flow, which is also called the Newtonian model. We will diverge a bit here, to talk about the behavior of materials as they flow under applied force and temperature.[10] We will begin to discuss the effect of the rate of strain on a material. Newton defined the relationship by using the dashpot as a model. An example of a dashpot would be a car's shock absorber or a French press coffee pot. These have a plunger header that is pierced with small holes through which the fluid is forced. Figure 2.11 shows the response of the dashpot model. Note that as the stress is applied, the material responds by slowly flowing through the holes. As the rate of the shear is increased, the rate of flow of the material also increases. For a Newtonian fluid, the stress–strain rate curve is a straight line, which can be described by the following equation:

$$\sigma = \eta\dot{\gamma} = \eta(\partial\gamma/\partial t) \tag{2.8}$$

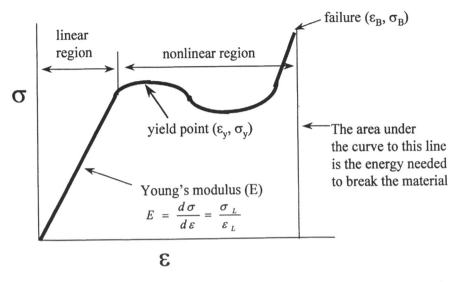

FIGURE 2.8 Dissecting a stress–strain curve. Analysis of a typical stress–strain curve in extension is shown. This is one of the most basic and most common tests done on solid materials.

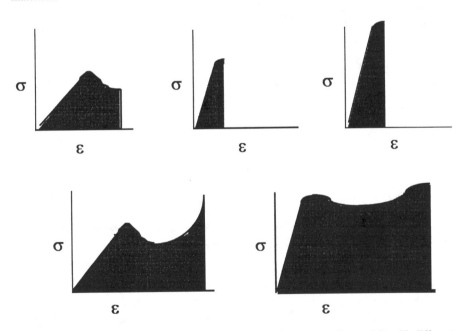

FIGURE 2.9 Stress-strain curves in extension for various types of materials with different mixtures of strength and toughness are shown. The area under the curve is often integrated to obtain the energy needed to break the sample and used as an indicator of toughness of the material.

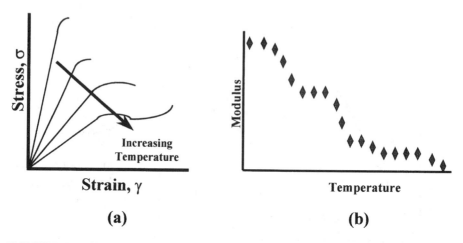

FIGURE 2.10 Changes as temperature increases. (a) Stress–strain curves change as the testing temperature increases. As a polymer is heated, it becomes less brittle and more ductile. (b) These data can be graphically displayed as a plot of the modulus vs. temperature.

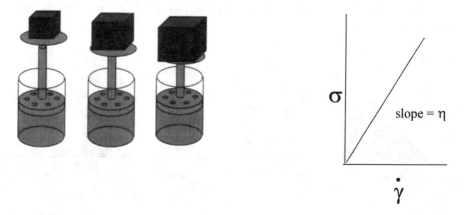

FIGURE 2.11 Newton's Law and dashpot. Flow is dependent on the rate of shear and there is no recovery seen. A dashpot, examples of which include a car's shock absorber or a French press coffee pot, acts as an example of flow or viscous response. The speed at which the fluid flows through the holes (the strain rate) increases with stress!

where stress is related to shear rate by the viscosity. This linear relationship is analogous to the stress–strain relation. While many oils and liquids are Newtonian fluids, polymers, food products, suspensions, and slurries are not.

The study of material flow is one of the largest areas of interest to rheologists, material scientists, chemists, and food scientists. Since most real materials are non-Newtonian, a lot of work has been done in this area. Non-Newtonian materials can be classified in several ways, depending on how they deviate from ideal behavior. These deviations are shown in Figure 2.12. The most common deviation is shear thinning. Almost all polymer melts are of this type. Shear thickening behavior is rare in polymer systems but often seen in suspensions. Yield stress behavior is also

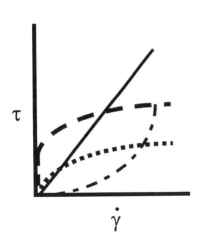

- Newtonian behavior is linear and the viscosity is independent of rate.
- Pseudoplastic (••••) fluids get thinner as shear increases.
- Dilatant Fluids (– – –) increase their viscosity as shear rates increase.
- Plastic Fluids (– – –) have a yield point with pseudoplastic behavior.
- Thixotrophic and rheopectic fluids (not shown) exhibit viscosity-time nonlinear behavior. For example, the former shear thin and then reform its gel structure.

FIGURE 2.12 Non-Newtonian behavior in solutions. The major departures from Newtonian behavior are shown in the figure.

observed in suspensions and slurries. Let's just consider a polymer melt, as shown in Figure 2.13, under a shearing force. Initially, a plateau region is seen at very low shear rates or frequency. This region is also called the zero-shear plateau. As the frequency (rate of shear) increases, the material becomes nonlinear and flows more. This continues until the frequency reaches a region where increases in shear rate no longer cause increased flow. This "infinite shear plateau" occurs at very high frequencies.

Like with solids, the behavior you see is dependent on how you strain the material. In shear and compression, we see a thinning or reduction in viscosity (Figure 2.14). If the melt is tested in extension, a thickening or increase in the viscosity of the polymer is observed. These trends are also seen in solid polymers. Both a polymer melt and a polymeric solid under frequency scans show a low-frequency Newtonian region before the shear thinning region. When polymers are tested by varying the shear rate, we run into four problems that have been the drivers for much of the research in rheology and complicate the life of polymer chemists.

These four problems are defined by C. Macosko[11] as (1) shear thinning of polymers, (2) normal forces under shear, (3) time dependence of materials, and (4) extensional thickening of melts. The first two can be solved by considering Hooke's Law and Newton's Law in their three-dimensional forms.[12] Time dependence can be addressed by linear viscoelastic theory. Extensional thickening is more difficult, and the reader is referred to one of several references on rheology[13] if more information is required.

2.5 ANOTHER LOOK AT THE STRESS–STRAIN CURVES

Before our discussion of flow, we were looking at a stress–strain curve. The curves of real polymeric materials are not perfectly linear, and a rate dependence is seen.

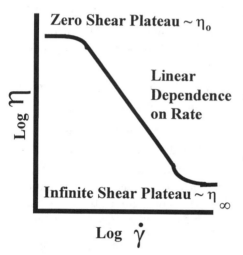

- At low shear rates, the viscosity is controlled by MW. The material shows Newtonian behavior
- Viscosity shows a linear dependence on rate above the η_o region.
- At high rates, the material no longer shows shear thinning and a second plateau is reached.

FIGURE 2.13 A polymer melt under various shear rates. Note Newtonian behavior is seen at very high and very low shear rates or frequencies. This is shown as a log-log-log plot, as is normally done by commercial thermal analysis software. Better ways of handling the data are now available. For example, the Carreau model described in Armstrong et al.[10] can be fitted using a regression software package like PolyMath or Mathematicia.

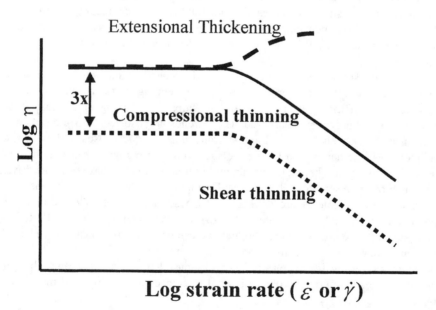

FIGURE 2.14 Shear, compressive, and extensional flows. While both compressive and shear cause an apparent thinning of the material, extensional flow causes a thickening. Note also that the modulus difference between shear and compression can be related as $1/3E = G$ for cases when Poisson's ratio, ν, is equal to 0.5.

Testing a Hookean material under different rates of loading shouldn't change the modulus. Yet, both curvature in the stress–strain curves and rate dependence are common enough in polymers for commercial computer programs to be sold that address these issues. Adding the Newtonian element to the Hookean spring gives a method of introducing flow into how a polymer responds to an increasing load (Figure 2.15). The curvature can be viewed as a function of the dashpot, where the material slips irrecoverably. As the amount of curvature increases, the increased curvature indicates the amount of liquid–like character in the material has increased. This is not to suggest that the Maxwell model, the parallel arrangement of a spring and a dashpot seen in Figure 2.15, is currently used to model a stress–strain curve. Better approaches exist. However, the introduction of curvature to the stress–strain curve comes from the viscoelastic nature of real polymers.

Several trends in polymer behavior[14] are summarized in Figure 2.16. Molecular weight and molecular weight distribution have, as expected, significant effects on the stress–strain curve. Above a critical molecular weight (M^c), which is where the material begins exhibiting polymer-like properties, mechanical properties increase with molecular weight. The dependence appears to correlate best with the weight average molecular in the Gel Permeation Chromatography (GPC). For thermosets, T_g here tracks with degree of cure. There is also a T_g value above which the corresponding increases in modulus are so small as to not be worth the cost of production. Distribution is important, as the width of the distributions often has significant effects on the mechanical properties.

In crystalline polymers, the degree of crystallinity may be more important than the molecular weight above the M^c. As crystallinity increases, both modulus and yield point increase, and elongation at failure decreases. Increasing the degree of crystallinity generally increases the modulus; however, the higher crystallinity can also make a material more brittle. In unoriented polymers, increased crystallinity can actually decrease the strength, whereas in oriented polymers, increased crystallinity and orientation of both crystalline and amphorous phases can greatly increase modulus and strength in the direction of the orientation. Side chain length causes increased toughness and elongation, but lowers modulus and strength as the length of the side chains increase. As density and crystallinity are linked to side chain length, these effects are often hard to separate.

As temperature increases we expect modulus will decrease, especially when the polymer moves through the glass transition (T_g) region. In contrast, elongation-to-break will often increase, and many times goes through a maximum near the midpoint of tensile strength. Tensile strength also decreases, but not to as great an extent as the elongation-to-break does. Modifying the polymer by drawing or inducing a heat set is also done to improve the performance of the polymer. A heat set is an orientation caused in the polymer by deforming it above its T_g and then cooling. This is what makes polymeric fibers feel more like natural fabrics instead of feeling like fishing line. The heat-set polymer will relax to an unstrained state when the heat-set temperature is exceeded. In fabrics, this relaxation causes a loss of the feel or "hand" of the material, so that knowing the heat set temperature is very important in the fiber industry. Cured thermosets, which can have decomposition temperature below the T_g, do not show this behavior to any great extent.

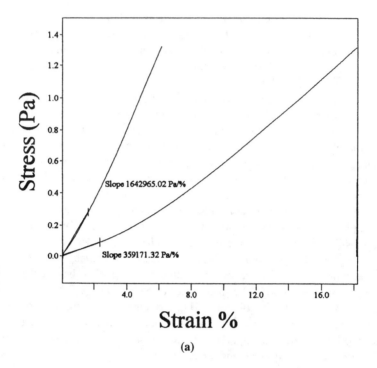

Slope 1642965.02 Pa/%

Slope 359171.32 Pa/%

Strain %

(a)

Curvature is caused by the
viscous part of the polymer,
represented by a dashpot.

(b)

FIGURE 2.15 The Maxwell model, an early attempt to explain material behavior, is a
dashpot and a spring in parallel. The dashpot introduces curvature into the graph, representing
the ability of the material to flow.

The elongation-to-break of thermosets is limited to a small percentage of its length
by the crosslinks present.

The addition of plasticizers to a polymer causes changes that look similar to
having increased the temperature, as shown in Figure 2.17. Plasticizers also cause
an increase the width of the T_g region. Fillers can act several ways, all with a

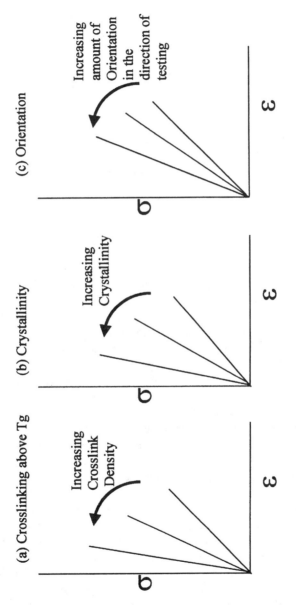

FIGURE 2.16 Effects of structural changes on stress–strain curves. As the structure of the polymer changes, certain changes are expected in the stress–strain curves.

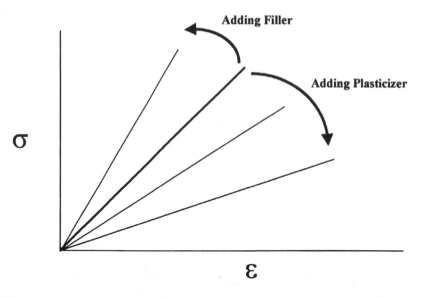

FIGURE 2.17 Plasticizers and fillers effects. Some fillers, specifically elastomers added to increase toughness and called tougheners, can also act to lower the modulus.

dependence on loading. Rigid fillers raise the modulus, while soft microscopic particles can lower the modulus while increasing toughness. The form of the filler is important, as powders will decrease elongation and ultimate strength as the amount of filler increases. Long fibers, on the other hand, cause an increase in both the modulus and the ultimate strength. In both cases, there is an upper limit to the amount of filler that can be used and still maintain the desired properties of the polymer. For example, if too high a weight percentage of fibers is used in a fiber-reinforced composite, there will not be enough polymer matrix to hold the composite together.

The speed of the application of the stress can show an effect on the modulus, and this is often a shock to people from a ceramics or metallurgical background. Because of the viscoelastic nature of polymers, one does not see the expected Hookean behavior where the modulus is independent of rate of testing. Increasing the rate causes the same effects one sees with decreased temperature: higher modulus, lower extension to break, less toughness (Figure 2.18). Rubbers and elastomers often are exceptions, as they can elongate more at high rates. In addition, removing the stress at the same rate it was applied will often give a different stress–strain curve than that obtained on the application of increasing stress (Figure 2.19). This hysteresis is also caused by the viscoelastic nature of polymers.

As mentioned above, polymer melts and fluids also show non-Newtonian behavior in their stress–strain curves. This is also seen in suspensions and colloids. One common behavior is the existence of a yield stress. This is a stress level below which one does not see flow in a predominately fluid material. This value is very important in industries such as food, paints, coating, and personal products (cosmetics, sham-

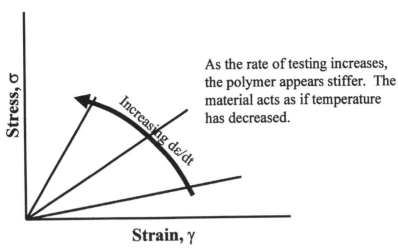

As the rate of testing increases, the polymer appears stiffer. The material acts as if temperature has decreased.

FIGURE 2.18 Rate of testing affects polymers because they are viscoelastic, not Hookean solids. Increasing the rate of applying stress acts as if we raised the temperature. All materials are at least slightly rate dependent, but the effect is small in elastic materials.

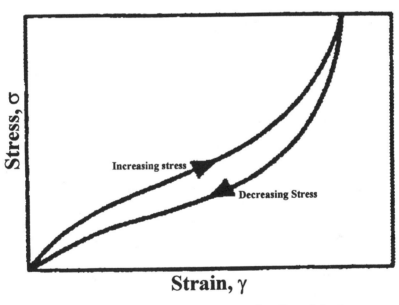

FIGURE 2.19 Hysteresis in polymers is an indicator of nonlinear behavior.

poo, etc.), where the material is designed to exhibit two different types of behavior. For example, mayonnaise is designed to spread easily when applied (one stress applied) but to cling to the food without dripping (another lower stress). The stress–strain curve for mayonnaise is shown in Figure 2.20a, where the knee or bend in the curve represents the value of the yield stress. The yield stress can also be

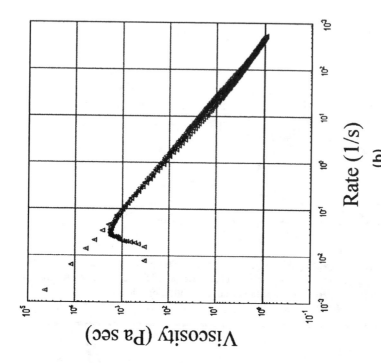

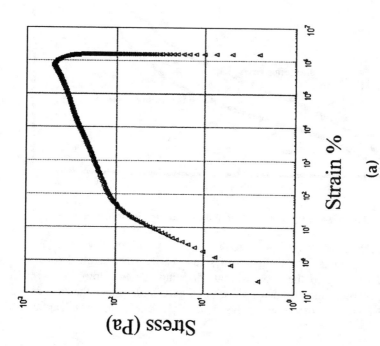

FIGURE 2.20   Stress–strain curves for mayonnaise. The yield stress in mayonnaise shows the affect of an important non-Newtonian behavior in food products. (a) A stress–strain curve with a visible knee at the yield stress. (b) Detecting the yield stress from a viscosity-rate plot.

determined from viscosity–shear rate curve, as shown in Figure 2.20b. Note the values don't agree. One needs to make sure the method used to determine the yield stress is a good representation of the actual use of the material.

None of these data are really useful for looking at how a polymer's properties depend on time. In order to start considering polymer relaxations, we need to consider creep–recovery and stress relaxation testing.

APPENDIX 2.1 CONVERSION FACTORS

Length

1 mil = 0.0000254 m
1 thou = 0.0254 mm
1 in. = 2.54 mm
1 ft = 304.8 mm
1 yd = 914.4 mm
1 mi = 1.61 km

Area

1 in.^{2} = 645.2 mm^2
1 ft^2 = 0.092 m^2
1 yd^2 = 0.8361 m^2
1 acre = 4047 m^2

Volume

1 oz. = 29.6 cm^3
1 in^3 = 16.4 cm^3
1 qt(l) = 0.946 dm^3
1 qt(s) = 1.1 dm^3
1 ft^3 = 0.028 dm^3
1 yd^3 = 0.0765 dm^3
1 gal(l) = 3.79 dm^3

Time

1 s = 9.19E-09 periods 55Cs133

Velocity

250 m/s = 55.9 mph
250 m/s = 90.6 kph
55 mph = 89.1 kph
55 mph = 245.9 m/s
90 kph = 55.6 mph

Acceleration

1 ft/s^2 = 0.3 m/s^2
1 free fall (g) = 9.806650 m/s^2

Frequency

1 cycle/s = 1 Hz
1 w = rad/s = 0.15915494 Hz
1 Hz = 6.283185429 w
1 Hz = 60.00 rpm
1 rpm = 0.1047198 r/s
1 rpm = 0.017 Hz

Plane Angle

1 degree = 0.017453293 rad
1 rad = 57.29577951 degree

Mass

1 carat (m) = 0.2 g
1 grain = 0.00000648 g
1 oz (av) = 28.35 g
1 oz (troy) = 31.1 g
1 lb = 0.4536 kg
1 ton (2000 lb) = 907.2 Mg

Force

1 dyne = 1.0000E-05 N
1 oz Force = 278 mN
1 g Force = 9.807 mN
1 mN = 0.101967982 g Force
1 lb Force = 4.4482E+00 N
1 ton Force (US) (2000 lb) = 8.896 kN
1 ton Force (UK) = 9.964 kN
1 ton (2000 lbf) = 8.8964E+03 N

Pressure

1 mm H_2O = 9.80E+00 Pa
1 lb/ft^2 = 4.79E+01 Pa
1 dyn/cm^2 = 1.00E+01 Pa
1 mmHg @ 0°C = 1.3332E+02 Pa
40 psi = 2.7579E+05 Pa 275790
300000 Pa = 4.3511E+01 psi 44
1 atm = 1.01E+05 Pa
1 torr = 1.33E+02 Pa
1 Pa = 7.5000E-03 torr
1 bar = 1.0000E+05 Pa
1 kPa = 1.00E+03 Pa
1 MPa = 1.00E+06 Pa
1 GPa = 1.00E+09 Pa
1 TPa = 1.00E+12 Pa

Viscosity (Dynamic)

  1 cP = 1.00E-03 Pa*s
  1 P = 1.00E-01 Pa*s
  1 kp*s/m^2 = 9.81E+00 Pa*s
  1 kp*h/m^2 = 3.53E+04 Pa*s

Viscosity (Kinematic)

  1 St = 1.00E-04 m^2/s
  1 cSt = 1.00E-06 m^2/s
  1 ft^2/s = 0.0929 m^2/s

Work (Energy)

  1 ft*lb = 1.36 J
  1 Btu = 1.05 J
  1 cal = 4.186 kJ
  1 kW*h = 3.6 MJ
  1 eV = 1.6E-19 J
  1 erg = 1.60E-07 J
  1 J = 0.73 ft*lbF
  1 J = 0.23 cal
  1 kJ = 1 Btu
  1 MJ = 0.28 kW*h

Power

  1 Btu/min = 17.58 W
  1 ft-lb/s = 1.4 W
  1 cal/s = 4.2 W
  1 hp (electric) = 0.746 kW
  1 W = 44.2 ft*lb/min
  1 W = 2.35 Btu/h
  1 kW = 1.34 hp (electric)
  1 kW = 0.28 ton (HVAC)

Temperature

  32°F = 491.7 R
  32°F = 0°C
  32°F = 273.2 K
   0°C = 32°F
   0°C = 273.2 K

NOTES

1. R. Steiner, *Physics Today,* 17, 62, 1969.
2. C. Macosko, *Rheology Principles, Measurements, and Applications,* VCH Publishers, New York, 1994.

3. Viscoelasticity is discussed later. For this chapter, we are assuming linear behavior unless we specifically state otherwise.

4. In this book, I will use the term *static stress* or *static force* to refer to the nonoscillatory stress of force applied to a sample to hold it in place. This is sometimes called the *clamping force*. The stress used in a creep example will be called a *constant stress* to indicate a constant load.

5. The addition of the dot over a symbol means we are using the rate of that property. In this case, the term could also be written as $\delta\sigma/\delta t$.

6. D. Holland, *J. Rheology,* 38(6), 1941, 1994. L. Kasehagen, *U. Minn Rheometry Short Course,* U. Minn., Minneapolis, 1996.

7. For a fuller development of Hookean behavior, see L. Nielsen et al., *Mechanical Properties of Polymers,* 3rd ed., Marcel Dekker, New York, 1994. S. Krishnamachari, *Applied Stress Analysis of Plastics,* Van Nostrand Reinhold, New York, 1993. C. Macosko, *Rheology Principles, Measurements, and Applications,* VCH Publishers, New York, 1994.

8. G. Gordon and M. Shaw, *Computer Programs for Rheologists,* Hanser Publishers, New York, 1994.

9. D. Askeland, *The Science and Engineering of Materials,* PWS Publishing, Boulder, CO, 1994.

10. For a more detailed discussion of flow, viscosity, and melt rheology, the following are suggested: J. Dealy et al., *Melt Rheology and Its Role in Polymer Processing,* Van Nostrand Reinhold, Toronto, 1990. J. Dealy, *Rheometers for Molten Plastics,* Van Nostrand Reinhold, Toronto, 1982. H. Barnes et al., *An Introduction to Rheology,* Elsevier, New York, 1989. N. Cheremisinoff, *An Introduction to Polymer Processing,* CRC Press, Boca Raton, FL, 1993. R. Tanner, *Engineering Rheology,* Oxford University Press, New York, 1988. R. Armstrong et al., *Dynamics of Polymer Fluids,* vol. 1 and 2, Wiley, New York, 1987. C. Rohn, *Analytical Polymer Rheology,* Hanser-Gardner, New York, 1995.

11. C. Macosko, *Rheological Measurements Short Course Text,* University of Minnesota, Minneapolis, 1996.

12. M. Tirrell and C. Macosko, *Rheological Measurements Short Course Text,* University of Minnesota, Minneapolis, 1996.

13. In addition to the references in 7 and 9 above, see J. D. Ferry, *Viscoelastic Properties of Polymers,* 3rd ed., Wiley, New York, 1980. J. J. Aklonis and W. J. McKnight, *Introduction to Polymer Viscoelasticity,* 2nd ed., Wiley, New York, 1983. L. C. E. Struik, *Physical Aging in Amorphous Polymers and Other Materials,* Elsevier, New York, 1978. W. Brostow and R. Corneliussen Eds., *Failure of Plastics,* Hanser, New York, 1986, Ch. 3. S. Matsuoka, Relaxation Phenomena in Polymers, Hanser, New York, 1992. N. McCrum, G. Williams, and B. Read, *Anelastic and Dielectric Effects in Polymeric Solids,* Dover, New York, 1994. M. Doi and S. Edwards, *The Dynamics of Polymer Chains,* Oxford University Press, New York, 1986.

14. The following discusion was extracted from several books. The best summaries are found in L. Nielsen et al., *Mechanical Properties of Polymers,* 3rd ed., Marcel Dekker, New York, 1994. C. Rohn, *Analytical Polymer Rheology,* Hanser-Gardner, New York, 1995. R. Seymour et al., *Structure Property Relationships in Polymers,* Plenum Press, New York, 1984. D. Van Krevelen, *Properties of Polymer,* 2nd ed., Elsevier, New York, 1990.

3 Rheology Basics: Creep–Recovery and Stress Relaxation

The next area we will review before starting on dynamic testing is creep, recovery, and stress relaxation testing. Creep testing is a basic probe of polymer relaxations and a fundamental form of polymer behavior. It has been said that while creep in metals is a failure mode that implies poor design, in polymers it is a fact of life.[1] The importance of creep can be seen by the number of courses dedicated to it in mechanical engineering curriculums as well as the collections of data available from technical societies.[2]

Creep testing involves loading a sample with a set weight and watching the strain change over time. Recovery tests look at how the material relaxes once the load is removed. The tests can be done separately but are most useful together. Stress relaxation is the inverse of creep: a sample is held at a set length and the force it generates is measured. These are shown schematically in Figure 3.1. In the following sections we will discuss the creep–recovery and stress relaxation tests as well as their applications. This will give us an introduction to how polymers relax and recover. As most commercial DMAs will perform creep tests, it will also give us another tool to examine material behavior.

Creep and creep–recovery tests are especially useful for studying materials under very low shear rates or frequencies, under long test times, or under real use conditions. Since the creep–recovery cycle can be repeated multiple times and the temperature varies independently of the stress, it is possible to mimic real–life conditions fairly accurately. This is done for everything from rubbers to hair coated with hairspray to the wheels on a desk chair.

3.1 CREEP–RECOVERY TESTING

If a constant static load is applied to a sample, for example, a 5-lb weight is put on top of a gallon milk container, the material will obviously distort. After an initial change, the material will reach a constant rate of change that can be plotted against time (Figure 3.2). This is actually how a lot of creep tests are done, and it is still common to find polymer manufacturers with a room full of parts under load that are being watched. This checks not only the polymer but also the design of the part.

More accurately representative samples of polymer can be tested for creep. The sample is loaded with a very low stress level, just enough to hold it in place, and allowed to stabilize. The testing stress is than applied very quickly, with instantaneous application being ideal, and the changes in the material response are recorded

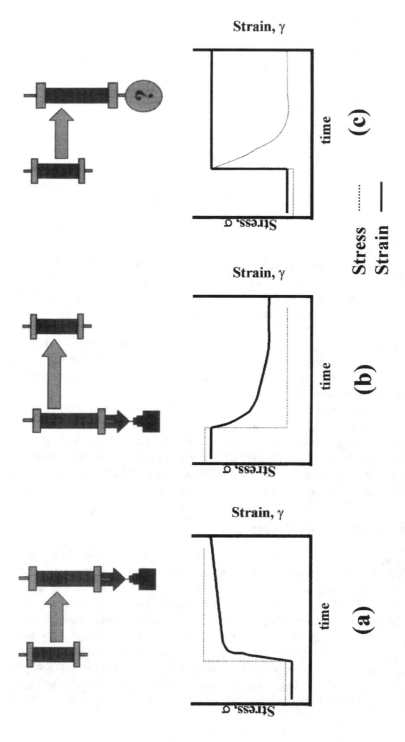

FIGURE 3.1 Creep, recovery, and stress relaxation tests. (a) Creep testing is performed by applying a load or stress to a sample. (b) When the stress is removed and the material allowed to recover, this is called a recovery test. These two tests are often cycled. (c) Stress relaxation is the reverse of creep. Holding a sample at a set length, the change in stress as a function of time or temperature is recorded.

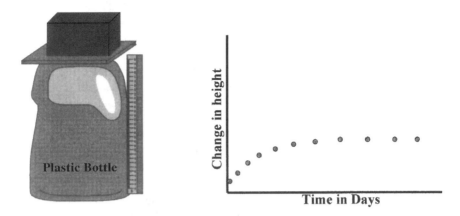

FIGURE 3.2 Creep tests. A representation of a simple creep test where the specimen is loaded with a stress that it would see in real life and its deformation tracked as a function of time.

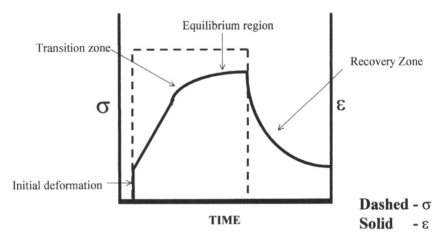

FIGURE 3.3 A creep-recovery curve as seen from a modern DMA operating in the creep-recovery mode showing (a) the applied stress curve and (b) the resulting strain curve. Note the idealized strain curve shows distinct regions of behavior related to (1) the initial deformation, (2) a transition zone, (3) the equilibrium region, and (4) the recovery region.

as percent strain. The material is then held at this stress for a period of time until the material reaches equilibrium. Figure 3.3 shows a creep test and the recovery step.

We can use creep tests in two ways: to gain basic information about the polymer or to examine the polymer response under conditions that approximate real use. In the former case, we want to work within the linear viscoelastic region so we can calculate equilibrium values for viscosity (η), modulus (E), and compliance (J). Compliance is the willingness of the material to deform and is the inverse of modulus for an elastic material. However, for a viscoelastic material this is not true, and a Lapace transform is necessary to make that conversion.[3] As we mentioned in the previous chapter, polymers have a range over which the viscoelastic properties are linear. We can determine this region for creep–recovery by running a series of tests on different specimens from the sample and plotting the creep compliance, J, versus time, t.[4] Where the plots begin to overlay, this is the linear viscoelastic region. Another approach to finding the linear region is to run a series of creep tests and observe under what stress no flow occurs in the equilibrium region over time (Figure 3.4). A third way to estimate the linear region is to run the curve at two stresses and add the curves together, using the Boltzmann superposition principle, which states that the effect of stresses is additive in the linear region. So if we look at the 25 mN curve in Figure 3.4 and take the strain at 0.5 min, we notice the strain increases linearly with the stress until about 100 mN, where it starts to diverge, and at 250 mN the strains are no longer linear. Once we have determined the linear region, we can run our samples within it and analyze the curve. This does not mean you cannot get very useful data outside this limit, but we will discuss that later.

Creep experiments can be performed in a variety of geometries, depending on the sample, its modulus and /or viscosity, and the mode of deformation that it would be expected to see in use. Shear, flexure, compression, and extension are all used. The extension or tensile geometry will be used for the rest of this discussion unless otherwise noted. When discussing viscosity, it will be useful to assume that the extensional or tensile viscosity is three times that of shear viscosity for the same sample when Poisson's ratio, v, is equal to 0.5.[5] For other values of Poisson's ratio, this does not hold.

3.2 MODELS TO DESCRIBE CREEP–RECOVERY BEHAVIOR

In the preceding chapter, we discussed how the dashpot and the spring are combined to model the viscous and elastic portions of a stress–strain curve. The creep–recovery curve can also be looked at as a combination of springs (elastic sections) and dashpots (viscous sections).[6] However, the models discussed in the last chapter are not adequate for this. The Maxwell model, with the spring and dashpot in series (Figure 3.5a) gives a strain curve with sharp corners where regions change. It also continues to deform as long as it is stressed for the dashpot continues to respond. So despite the fact the Maxwell model works reasonably well as a representation of stress–strain curves, it is inadequate for creep.

The Voigt–Kelvin model with the spring and the dashpot in parallel is the next simplest arrangement we could consider. This model, shown in Figure 3.5b, gives a curve somewhat like the creep–recovery curve of a solid. This arrangement of the spring and dashpot gives us a way to visualize a time-dependent response as the resistance of the dashpot slows the restoring force of the spring. However, it doesn't show the instantaneous response seen in some samples. It also doesn't show the continued flow under equilibrium stress that is seen in many polymers.

In order to address these problems, we can continue the combination of dashpots and springs to develop the four-element model. This combining of the various dashpots and spring is used with fair success to model linear behavior.[7] Figure 3.5c

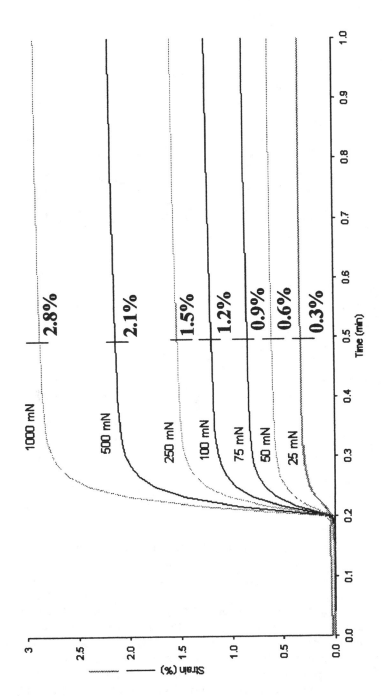

FIGURE 3.4 Linear region from creep–response. A plot of percent strain against time showing two methods of determining the linear region for a creep curve. One can look for the region where the equilibrium region shows no flow as a function of time or where the stress-strain relationship ceases being linear.

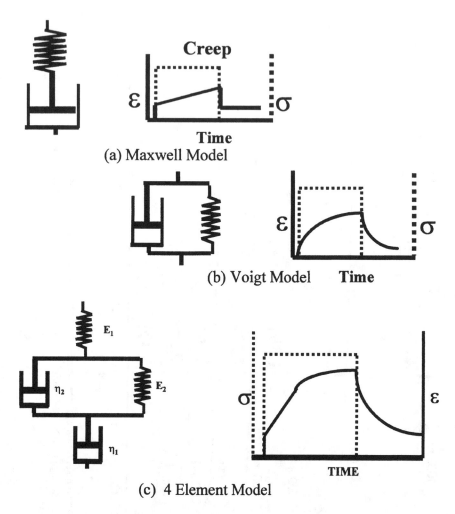

FIGURE 3.5 Models for approximating creep–recovery response. Neither the Maxwell (a) nor the Voigt (b) model work well to explain creep. The four-element model (c) does a better job.

shows the model and the curve that results from it. This curve shows the same regions as seen in real materials, including a small instantaneous region, a leveling off of the equilibrium region, and a realistic recovery curve. We can use the four-element model to help us understand the strain curve. We can also add additional elements if needed to adjust the behavior and tie it back to structural units. This is a common approach, and Shoemaker et al.[8] report the use of a six-element model to predict the behavior of ice cream where various parts of that mixture were assigned to specific elements of the model. For example, they assign the independent spring to ice crystals, the independent dashpot to butterfat, and the Voigt elements to stabilizer gels, air cells, and fat crystals. This approach doesn't always work this well (and there are some doubts as to the validity of these assignments in this particular case too[9]), and better approaches exist. While real polymers do not have springs and dashpots in them, the idea gives us an easy way to explain what is happening in a creep experiment.

3.3 ANALYZING A CREEP–RECOVERY CURVE TO FIT THE FOUR-ELEMENT MODEL

If we now examine a creep–recovery curve, we have three options in interpreting the results. These are shown graphically in Figure 3.6. We can plot strain vs. stress and fit the data to a model, in this case to the four-element model as shown in Figure 3.6. Alternately, we could plot strain vs. stress and analyze quantitatively in terms of irrecoverable creep, viscosity, modulus, and relaxation time. A third choice would be to plot creep compliance, J, versus time.

In Figure 3.6a, we show the relationship of the resultant strain curve to the parts of the four-element model. This analysis is valid for materials in their linear viscoelastic region and only those that fit the model. However, it is a simple way to separate sample behavior into elastic, viscous, and viscoelastic components. As the stress, σ_o, is applied, there is an immediate response by the material. The point at which σ_o is applied is when time is equal to 0 for the creep experiment. (Likewise for the recovery portion, time zero is when the force is removed.) The height of this initial jump is equal to the applied stress, σ_o, divided by the independent spring constant, E_1. This spring can be envisioned as stretching immediately and then locking into its extended condition. In practice, this region may be very small and hard to see, and the derivative of strain may be used to locate it. After this spring is extended, the independent dashpot and the Voigt element can respond. When the force is removed, there is an immediate recovery of this spring that is again equal to σ_o/E_1. This is useful, as sometimes it is easier to measure this value in recovery than in creep. From a molecular perspective, we can look at this as the elastic deformation of the polymer chains.

The independent dashpots contribution, η_1, can be calculated by the slope of the strain curve when it reaches region of equilibrium flow. This equilibrium slope is equal to the applied stress, σ_o, divided by η_1. The same value can be obtained determining the permanent set of the sample, and extrapolating this back to t_f, the time at which σ_o was removed. A straight line drawn for t_o to this point will have

$$\text{slope} = \sigma_o(t_f)/\eta_1 \tag{3.1}$$

The problem with this method is that the time required to reach the equilibrium value for the permanent set may be very long. If you can actually reach the true permanent set point, you could also calculate η_1 from the value of the permanent set directly. This dashpot doesn't recover because there is nothing to apply a restoring force to it, and molecularly it represents the slip of one polymer chain past another.

The curved region between the initial elastic response and the equilibrium flow response is described by the Voigt element of the Berger model. Separating this into individual components is much trickier, as the region of the retarded elastic response

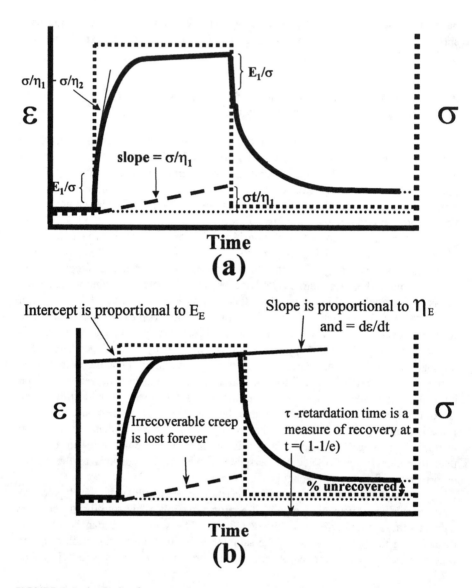

FIGURE 3.6 Analysis of a creep–recovery curve when viewed as either strain and compliance vs. time. (a) The black terms correspond to the parts of the four-element model discussed in the chapter. (b) The displayed values are the most commonly used ones in the author's experience.

is described by the parallel combination of the spring, E_2, and dashpot, η_2. In addition, some contribution from the independent dashpot exists. This region responds slowly due the damping affect of the dashpot until the spring is fully extended. The presence of the spring allows for a slow recovery as it pushes the dashpot back to its original position. Molecularly, we can consider this dashpot to represent the resistance of the chains to uncoiling, while the spring represents the thermal vibration of chain segments that will tend to seek the lowest energy arrangement.

Since the overall deformation of the model is given as

$$\varepsilon(f) = \left(\sigma_o/E_1\right) + \left(\sigma_o/\eta_1\right) + \left(\sigma_o/E_2\right)\left(1 - e^{-t/(\eta_2/E_2)}\right) \qquad (3.2)$$

we can get the value for the Voigt unit by subtracting the first two terms from the total strain, so

$$\varepsilon(f) - \left(\sigma_o/E_1\right) - \left(\sigma_o/\eta_1\right) = \left(\sigma_o/E_2\right)\left(1 - e^{-t/(\eta_2/E_2)}\right) \qquad (3.3)$$

The exponential term, η_2/E_2, is the retardation time, τ, for the polymer. The retardation time is the time required for the Voigt element to deform to 63.21% (or $1 - 1/e$) of its total deformation. If we plot the log of strain against the log of time, the creep curve appears sigmoidal, and the steepest part of the curve occurs at the retardation time. Taking the derivative of the above curve puts the retardation time at the peak. Having the retardation time, we can now solve the above equation for E_2 and then get η_2. The major failing of this model is it uses a single retardation time when real polymers, due to their molecular weight distribution, have a range of retardation times.

A single retardation time means this model doesn't fit most polymers well, but it allows for a quick, simple estimate of how changes in formulation or structure can affect behavior. Much more exact models exist,[10] including four-element models in 3D and with multiple relaxation times, but these tend to be mathematically nontrivial. A good introduction to fitting the models to data and to multiple relaxation times can be found in Sperling's book.[11]

3.4 ANALYZING A CREEP EXPERIMENT FOR PRACTICAL USE

The second of the three methods of analysis, shown in Figure 3.6b, is more suited to the real world. Often we intentionally study a polymer outside of the linear region because that is where we plan to use it. More often, we are working with a system that does not obey the Berger model. If we look at Figure 3.6, we can see that the slope of the equilibrium region of the creep curve gives us a strain rate, $\dot{\varepsilon}$. We can also calculate the initial strain, ε_o, and the recoverable strain, ε_r. Since we know the stress and strain for each point on the curve we can calculate a modulus (σ/ε) and, with the strain rate, a viscosity ($\sigma/\dot{\varepsilon}$). If we do the latter where the strain rate has become constant, we can measure an equilibrium viscosity, η_e. Extrapolating that line back to t_o, we can calculate the equilibrium modulus, E_e. Percent recovery and a relaxation time can also be calculated. These values help quantify the recovery cure: percent recovery is simply how much the polymer comes back after the stress is released, while the relaxation time here is simply the amount of time required for the strain to recover to 36.79% (or $1/e$) of its original value.

We can actually measure three types of viscosity from this curve. The simple viscosity is given above, and by multiplying the denominator by 3 we approximate the shear viscosity, η_s. Nielsen suggests that a more accurate viscosity, $\eta_{\Delta\varepsilon}$, can be obtained by inverting the recovery curve and subtracting it from the creep curve. The resulting value, $\Delta\varepsilon$, is then used to calculate a strain rate, multiplied by 3 and divided into the stress, σ_0. Finally we can calculate the irrecoverable viscosity, η_{irr}, by extrapolating the strain at permanent set back to t_f and taking the slope of the line from t_0 to t_f. This slope can be used to calculate an irrecoverable strain rate, which is then multiplied by 3 and divided into the initial stress, σ_0. This value tells us how quickly the material flows irreversibly.

If we instead choose to plot creep compliance against time, we can calculate various compliance values. Extrapolating the slope of the equilibrium region back to t_0 gives us J_e^0, while the slope of this region is equal to t/η_0. The very low shear rates seen in creep, this term reduces to $1/\eta_0$. We can also use the recovery curve to independently calculate J_e^0 by allowing the polymer to recover to equilibrium. Since we know

$$\lim_{t \to \infty} J_r(t) = J_e^0 \quad \text{for} \quad \dot{\varepsilon}(t) = \dot{\varepsilon}_\infty \tag{3.4}$$

then we can watch the change in $\dot{\varepsilon}$ until it is zero or, more practically, very small. This can be done by watching the second derivative of the strain as it approaches zero. At this point, J_r is equal to J_e^0. If we are in steady state creep, the two measurements of J_e^0 should agree. If we actually measure the J_e^0, we can estimate the longest retardation time (λ_0) for the material by $\eta_0 * J_e^0$.

3.5   OTHER VARIATIONS ON CREEP TESTS

Before we discuss the structure–property relationships or concepts of retardation and relaxation times, lets quickly look at variations of the simple creep–recovery cycle we discussed above. As we said before, a big advantage of a creep test is its ability to mimic the conditions seen in use. By varying the number cycles and the temperature, we can impose stresses that approximate many end-use conditions.

Figure 3.7 shows three types of tests that are done to simulate real applications of polymers. In Figure 3.7a, multiple creep cycles are applied to a sample. This can be done for a set number of cycles to see if the properties degrade over multiple cycles (for example, to test a windshield wiper blade) or until failure (for example on a resealable o-ring). Creep testing to failure is also occasionally called a creep rupture experiment. One normally analyzes the first and last cycle to see the degree of degradation or plots a certain value, say η_e, as a function of cycle number.

You can also vary the temperature with each cycle to see where the properties degrade as temperature increases. This is shown in Figure 3.7b. The temperature can be raised and lowered, to simulate the effect of an environmental thermal cycle. It can also be just raised or lowered to duplicate the temperature changes caused by

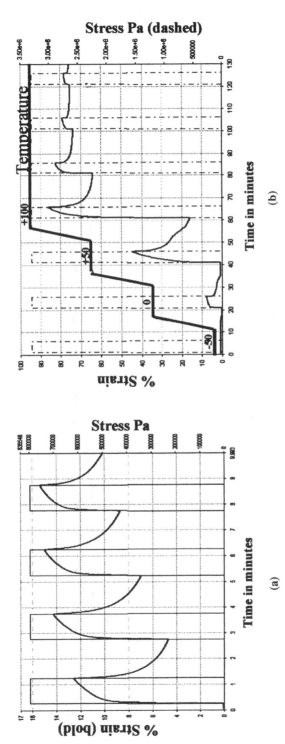

FIGURE 3.7 Examples of types of creep tests: (a) Multiple creep–recovery cycles, (b) multiple creep cycles with overlaying temperature ramp, and (c) heat-set cycle.

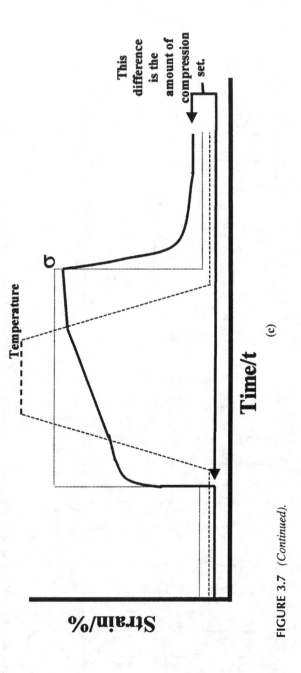

FIGURE 3.7 (*Continued*).

placing the part into a specific environment, such as a gasket in a pump down an oil well or a plastic pipe in an Alaskan winter. This environmental testing is not limited to temperature, as creep–recovery tests can also be run in solvents or in controlled atmospheres.

You can also vary the temperature within one creep cycle, as shown in Figure 3.7c. This is the equivalent of the rubber industry's heat-set test, used for materials that will be heated and squeezed at the same time. The creep stress is applied and the material is heated to a set temperature and cooled back to room temperature while still under the load. The stress is removed and the amount of recovery recorded.

A final comment on creep testing is that the American Society for Testing and Materials (ASTM)[12] does have standard procedures for creep tests that supply guidelines for both testing and data interpretation. The main method for plastics is D 2990-91. It covers tensile, compressive, and flexural creep and creep rupture.

3.6 A QUICK LOOK AT STRESS RELAXATION EXPERIMENTS

The conceptual inverse of a creep experiment is a stress relaxation experiment (Figure 3.8a). A sample is very quickly distorted to a set length, and the decay of the stress exerted by the sample is measured. These are often difficult experiments to run, because the sample may need to be strained very quickly. They do provide some very useful information that complements creep data. Creep data and stress relaxation data can be treated as mainly reciprocal,[13] and roughly related as

$$\left(\varepsilon_t / \varepsilon_o\right)_{creep} \approx \left(\sigma_o / \sigma_t\right)_{stress\ relax} \tag{3.5}$$

The analysis of the stress relaxation curve is shown in Figure 3.8a and is analogous to the creep analysis. One interesting application of the stress relaxation experiment exploits the relationship that the area under the stress relaxation curve plotted as $E(t)$ versus t is the viscosity, η_o. Doing experiments at very low strain, this allows us to measure the viscosity of a colloid without destroying its structure.[14]

A special type of stress relaxation experiment with immense industry applications is the constant gauge length experiment.[15] This is shown in Figure 3.8b. A sample is held at a set length with a minimal stress and then the temperature is increased. As the material responds to the temperature changes, the stress exerted by it is measured. The shrinkage or expansion force of the material is recorded. The experiment may be done with thermal cycles to determine if the same behavior is seen during each cycle.

3.7 SUPERPOSITION — THE BOLTZMANN PRINCIPLE

The question sometimes arises of how strains act when applied to a material that is already deformed. Boltzmann showed back in 1876 that the strains will add together linearly and a material's stress at any one time is a function of its strain history.

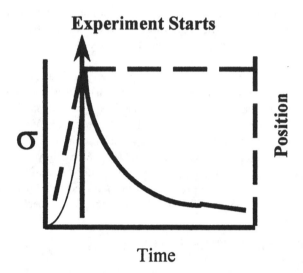

Experiment Starts

σ

Position

Time

Sample would be distorted to length y' and held.

(a) Classical Test

Force (mN)

550.0
500.0
450.0
400.0
350.0
300.0
250.0
200.0
150.0
100.0
50.0
0.0

25.0   50.0   75.0   100.0   125.0   150.0

Temperature (C)

(b) Constant Gauge Length

FIGURE 3.8 Stress relaxation experiments: (a) Analysis of a classical stress relaxation experiment where the sample is held at length, *l*, and the stress changes are recorded, (b) a constant gauge length experiment where the sample is held at length, *l*, and the temperature increased.

This applies to a linear response, no matter whether any of the models we discussed is applied. It also works for applied stress and measured strain. There is time dependence in this, as the material will change over time. For example, in stress relaxation the sample will have decreasing stress with time, and therefore in calculating the sum of the strains one needs to consider this decay to correctly determine the stress. This decay over time is called a memory function.[16]

The superposition of polymer properties is not just limited to the stress and strain effects. Creep and stress relaxation curves collected at different temperatures are also superpositioned to extend the range of data at the reference temperature. This will be discussed in detail in Chapter 8.

3.8　RETARDATION AND RELAXATION TIMES

We mentioned in Section 3.4 that one of the failings of the four-element model is that it uses a single retardation where most polymers have a distribution of retardation times. We also mentioned that we could estimate the longest retardation time from the creep compliance, J, versus time plot. The distribution of retardation times, $L(t)$, in a creep experiment or of relaxation times, $H(t)$, in a stress relaxation experiment are what determines the mechanical properties of a polymer. One method estimates $L(t)$ from the slope of the compliance curve against log (time) plot, and $H(t)$ is similarly obtained from the stress relaxation data. Below T_g, these are heavily influenced by the free volume, v_f, of the material. There is considerable interest in determining what the distribution of relaxation or retardation times are for a polymer, and many approaches can be found.[17] Again, Ferry[10] remains the major lead reference for those interested in this topic.

If you know the retardation time or relaxation time spectra, it is theoretically possible to calculate other types of viscoelastic data. This has not reduced to practice as well as one might hope, and the calculations are very complex. Neither $L(t)$ nor $H(t)$ are routinely used in solving problems. Methods also exist of calculating a discrete spectrum of relaxation and retardation times.[18]

3.9　STRUCTURE–PROPERTY RELATIONSHIPS IN CREEP–RECOVERY TESTS

The effects of various structural and environmental parameters on creep–recovery tests are well known.[19] Temperature may be the most important variable, as most materials show markedly different behavior above and below T_g (Figure 3.9a). The glass transition, T_g, of a polymer, where the polymer changes from glassy to rubbery, is where chains gain enough mobility to slide by each other. Below the T_g, the behavior of the polymer is dominated by the free volume, v_f, which limits the ability of the chains to move. In glassy polymers below the T_g where little molecular motion occurs, the amount of creep is small until the deformation is great enough to cause crazing.[20] Decreasing the ability of the chains to move, by lowering the temperature, increasing the pressure, annealing, increasing the degree of crystallinity, increasing the amount of cross-linking, or decreasing the free volume will decrease the amount of creep.

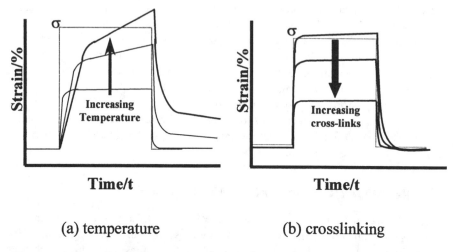

(a) temperature   (b) crosslinking

FIGURE 3.9 **Various effects from material changes on creep** showing (a) the effect of temperature and (b) the effect of cross-linking on the strain curve.

As the polymer temperature approaches the temperature of its glass transition, the amount of creep becomes very temperature-dependent. As we exceed the T_g, the affects of other structural parameter appear. Amorphous polymers without cross-linking have a strong dependence on molecular weight in the amount of flow seen at equilibrium. Branching changes the amount of creep, but the effect is difficult to summarize as depending on the branch length and degree of branching.[21] For example, the amount of flow can be either increased or decreased depending on whether the branches are long enough to entangle.[22] Plasticizers increase the creep. Plasticization acts by lowering the T_g, increasing the molecular weight between entanglements, and diluting the polymer. These effects also decrease the recovery.[23] Time can also affect recovery as long creep experiments allow more chains to disentangle and slip, lowering recovery.

Cross-linked polymers show a very specific curve with a flat equilibrium region, because the cross-links do not allow flow. As shown in Figure 3.9b, recovery is normally quite high. While some creep does occur if times are long enough and the cross-link density low enough, in highly cross-linked materials no creep is seen. For highly crystalline polymers below T_g, we see creep responses similar to those seen in a cross-linked polymer because the crystals act as cross-links and restrain flow. As crystallinity decreases, the material becomes less rigid. However, these materials are very sensitive to thermal history between the T_g and the melt because of changes to the crystal morphology.

3.10 THERMOMECHANICAL ANALYSIS

Thermomechanical analysis (TMA)[24] involves applying a static force to a sample and watching the dimensional changes. In some cases, as in testing a sample in flexure or compression, it could be viewed as a creep test with varying temperature.

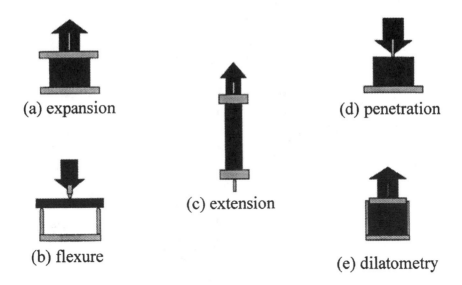

FIGURE 3.10 Thermomechanical analysis test methods in common use. Dilatometry allows one to measure volumetric expansion: all other tests measure only unidirectional change.

A more common use of TMA is the measure of the thermal expansion of a specimen. Since the basis of TMA's operation is the change in the dimensions of a sample as a function of temperature or time, a simple way of looking at TMA is as a very sensitive micrometer. Several testing approaches exist and are shown in Figure 3.10. Although it is not normally discussed in a development of DMA, its dependence on the same free volume effects that lead to relaxation and retardation times indicate that a discussion of the basics of this technique should be included here.

The technique of TMA was probably developed from penetration and hardness tests and first applied to polymers in 1948.[25] It is still a very commonly used method. As discussed above, the T_g in a polymer corresponds to the expansion of the free volume, allowing greater chain mobility above this transition. We will discuss thermal transitions at length in Chapter 5, but we need to mention here that the transitions are caused by increases in the free volume of the material as it is heated (Figure 3.11a). Seen as an inflection or bend in the thermal expansion curve, this change in the TMA can be seen to cover a range of temperatures (Figure 3.11b), of which the T_g value is an indicator defined by a standard method.[26] This fact seems to get lost as inexperienced users often worry why perfect agreement isn't seen in the value of the T_g when comparing different methods. The width of the T_g transition can be as important an indicator of changes in the material as the actual temperature. Other commonly used TMA methods such as flexure (Figure 3.11c) and extension (Figure 3.11d) also show that the T_g occurs over a range of temperatures. This is also seen in the DMA data (Chapter 5) as well as in the DSC.[27]

TMA allows the calculation of the coefficient of thermal expansion (CTE) or, more correctly, the thermal expansivity from the same data as used to determine the T_g (Figure 3.11b). Since many materials are used in contact with a dissimilar material

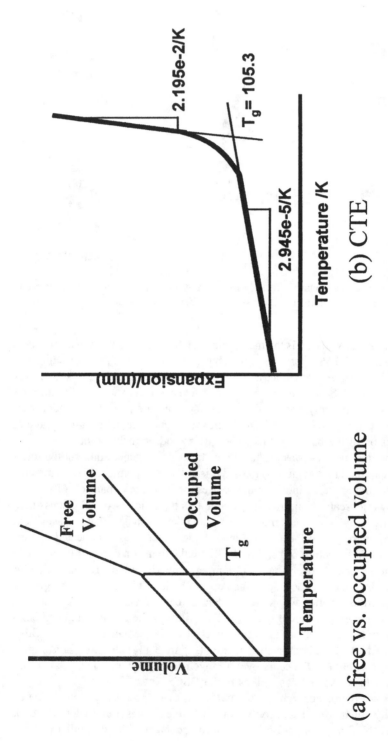

(a) free vs. occupied volume

(b) CTE

FIGURE 3.11 Transitions in TMA: (a) Changes occur in the free volume, (b) standard method for measuring the T_g by the CTE in expansion (note that the T_g occurs over a range of temperatures and not at a specific temperature, although the method picks a temperature as an indicator), (c) the T_g can be measured in penetration, (d) the T_g and heat-set temperatures are shown in extension. A flexure cure is not shown but would look similar to the curve in (c).

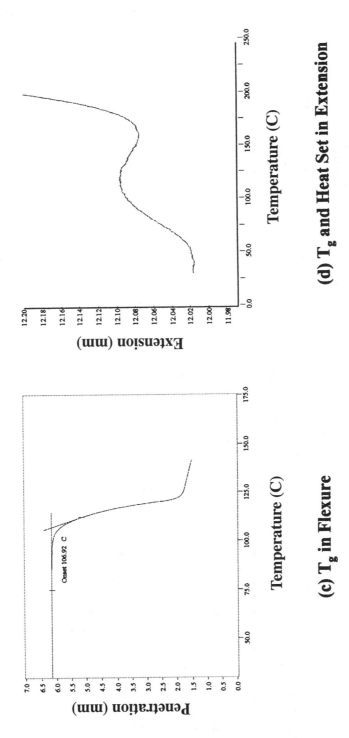

(d) T$_g$ and Heat Set in Extension

(c) T$_g$ in Flexure

FIGURE 3.11 (Continued).

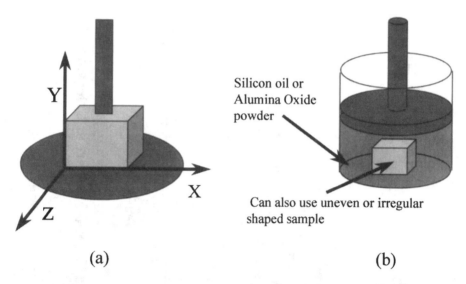

FIGURE 3.12 Anisotropic samples. Two approaches to measuring the expansion of an anisotropic material for a cubic sample: (a) Measuring each axis independently, and (b) measuring a bulk expansion. The latter will also work for irregularly shaped samples.

in the final product, knowing the rate and amount of thermal expansion by the CTE helps design around mismatches that can cause failure in the final product. These data are only available when the T_g is collected by thermal expansion in a technique formally called thermodilatometry.[28] Different T_g values are seen for each mode of testing (flexure, penetration, or expansion), as they each measure a slightly different effect.[29] It is also important to remember that anisotropic materials will have different CTEs depending on the direction in which they have been measured (Figure 3.12a). For example, a composite of graphite fibers and epoxy will show three distinct CTEs corresponding to the x-, y-, and z-directions, because the fibers have different orientation or packing in each axis. Blends of liquid crystals and polyesters show a significant enough difference between directions that the orientation of the crystals can be determined by TMA.[30] Similarly, oriented fibers and films show a different CTE in the direction of orientation.

The study of bulk or volumetric expansion is referred to as dilatometry and can also be done in a TMA. This technique, which involves measuring the volumetric change of a sample, is traditionally applied to liquids. Boundy and Boyer[31] used it extensively to measure initial rates for bulk polymerization of styrene. Because of the great difference in density between a polymer and its monomer, very accurate measurements are possible at very low degrees of conversion. When a total change in size is needed, for example in anisotropic or heterogeneous samples or in a very odd-shaped sample, a similar technique is used with silicon oil or alumina oxide as a filler. This is shown in Figure 3.12b. Volume changes like this are especially important in curing studies where the specimen shrinks as it cures.[32] The shrinkage of a thermoset as it cross-links often leads to cracking and void formation, and measurement of the amount of shrinkage is often critical to understanding the process.

NOTES

1. J. Sosa, *University of Houston Short Course on Polymer Analysis,* Houston, TX, 5 May 1994.
2. W. Andrew, *The Effect of Creep and Other Time Related Factors on Plastics,* vol. 1 and 2, Plastic Design Library, New York, 1991.
3. M. Tirrell, *Linear Viscoelasticity* at the 21st Short Course on Rheometry, U. Minnesota, Minneapolis, 1996.
4. C. Macosko, *Rheology Principles, Measurements, and Applications,* VCH, New York, 1994, pp. 119–121. L. Kasehagen, *Constant Stress Experiments* at 21st Short Course on Rheometry, U. Minnesota, Minneapolis, 1996.
5. T. Fox et al., in *Rheology,* vol. 1, Eirich, F., Ed., Academic Press, New York, 1956.
6. For a more detailed discussion of these models and their use, please refer to C. Macosko, *Rheology Principles, Measurements, and Applications,* VCH, New York, 1994. L. Neilsen et al., *Mechanical Properties of Polymers and Composites,* 3rd ed., Marcel Dekker, New York, 1994. S. Rosen, *Fundamental Principles of Polymeric Materials,* Wiley, New York, 1993.
7. N. Tschoegl, *The Phenomenological Theory of Linear Viscoelasticity,* Springer-Verlag, Berlin, 1989.
8. C. Shoemaker et al., *Society of Rheology Short Course on Food Rheology,* S. Rheology, Boston, 1993.
9. J. Ellis, *Society of Rheology Short Course on Food Rheology,* S. Rheology, Boston, 1993.
10. J. Ferry, *Viscoelastic Properties of Polymers,* 3rd ed., Wiley, New York, 1980. Tschoegl, N., *The Phenomenological Theory of Linear Viscoelasticity,* Springer-Verlag, Berlin, 1989. C. Macosko, *Rheology Principles, Measurements, and Applications,* VCH, New York, 1994. J. Skrzypek, *Plasticity and Creep,* CRC Press, FL, Boca Raton, 1993.
11. L. H. Sperling, *Introduction to Physical Polymer Science,* 2nd ed., Wiley, New York, 1992, pp. 458–497.
12. The ASTM can be reached at ASTM Committee on Standards, 1916 Race Street, Philadelphia, PA, 19103. Standards are republished yearly.
13. L. Neilsen, *Mechanical Propeties of Polymers and Composites,* vol. 1, Marcel Dekker, New York, 1974.
14. M. Tirrell, *Linear Viscoelasticity* at the 21st Short Course on Rheometry, U. Minnesota, Minneapolis, 1996.
15. C. Daley, and K. Menard, *NATAS Notes,* 26(2), 56, 1994.
16. We are not going to address the Boltzmann principle in detail, as it is beyond the scope of this book. Interested readers are referred to Ferry, op. cit.
17. A. Tobolsky, *Properties and Structure of Polymers,* Wiley, New York, 1964. J. Honerkamp, and J. Weese, *Rheologia Acta,* 32, 65, 1993. N. Orbey, and J. Dealy, *J. Rheology,* 1991, 35(6), 1035. L. Neilsen et al., *Mechanical Properties of Polymers and Composites,* 3rd ed., Marcel Dekker, New York, 1994.
18. J. Kaschta, and F. Schwarzl, *Rheologia Acta,* 33, 17, 1994. J. Kaschta and F. Schwarzl, *Rheologia Acta,* 33, 530, 1994. I. Emri and N. Tschoegl, *Rheologia Acta,* 32, 311, 1993. I. Emri and N. Tschogel, *Rheologia Acta,* 33, 60, 1994.
19. Most data on creep are well known, although work continues on specialized problems and new systems. The following is summarized from J. Ferry, *Viscoelastic Properties of Polymers,* 3rd ed., Wiley, New York, 1980. N. Tschoegl, *The Phenomenological Theory of Linear Viscoelasticity,* Springer-Verlag, Berlin, 1989. C. Macosko, *Rheol-*

ogy Principles, Measurements, and Applications, VCH, New York, 1994. J. Skrzypek, *Plasticity and Creep,* CRC Press, Boca Raton, FL, 1993. J. Mark et al., *Physicial Properties of Polymers,* ACS, Washington, D.C., 1984. N. McCrum et al., *Principles of Polymer Engineering,* Oxford University Press, New York, 1990. W. Brostow and R. Corneliussen, *Failure of Plastics,* Hanser-Gardiner, New York, 1989.

20. L. Nielsen and R. Landel, *Mechanical Properties of Polymers and Composites,* 2nd ed., Marcel Dekker, New York, 1994, pp. 89–120. J. Wu et al., *J. Rheology,* 23, 231, 1979. N. Brown et al., *J. Polym. Sci. (Phys.),* 16, 1085, 1978. C. Bucknall et al., *J. Mater. Sci.,* 7, 202, 1972.

21. F. Bueche, *J. Chem. Phys.,* 40, 484, 1964.

22. R. Chartoff and B. Maxwell, *J. Polymer Sci. A2,* 8, 455, 1970.

23. A. Tobolosky et al., *Polym. Eng. Sci.,* 10, 1, 1970.

24. A. Riga and M. Neag, *Material Characterization by Thermomechanical Analysis,* ASTM, Philadelphia, 1991.

25. V. Kargin et al., *Dokl. Akad. Nauk SSSR,* 62, 239, 1948.

26. For example see ASTM Method E 831, *Linear Thermal Expansion of Solids, in ASTM Standard Methods,* ASTM, Philadelphia, 1995, or H. McAdie, *Anal. Chem.,* 46, 1146, 1974.

27. K. Menard, in *Performance of Plastics,* W. Brostow, Ed., Hanser, New York, 1999, Ch. 7.

28. J. Hill, *For Better Thermal Analysis and Calorimetry,* ITHAC, New South Wales, Australia, 1991.

29. G. Curran, J. Rogers, H. O'Neal, S. Welch, K. Menard, *J. Advanced Materials,* 26(3), 49, 1995.

30. W. Brostow et al., paper in review.

31. R. Boundy and R. Boyer, *Styrene: Its Polymers, Copolymers, and Derivatives,* Reinhold, New York, 1952.

32. B. Bilyeu, W. Brostow, and K. Menard, *Proceedings of the 6th International Conference on Polymer Characterization,* Denton, TX, 1998, 67.

4 Dynamic Testing

In this chapter, we will address the use of a dynamic force to deform a sample. We have already looked at how a polymer exhibits both elastic (spring-like) and viscous (dashpot-like) behavior and how combination of these elements allows us to devise simple models of polymer behavior. We have seen that polymers have a time-dependent form of behavior and a "memory." Finally, we have seen that the free volume of polymers is a function of temperature.

In this chapter we shall begin by discussing the application of a dynamic force to a polymeric material. We shall then consider what happens if we increase the force, as in a stress–strain curve. Finally, we will take a brief look at instrumentation and fixtures. In the following chapters, we will address the question of scanning temperature (Chapters 5 and 6) and varying frequency (Chapter 7).

4.1 APPLYING A DYNAMIC STRESS TO A SAMPLE

If we take a sample at constant load and start sinusoidally oscillating the applied stress (Figure 4.1), the sample will deform sinusoidally. This will be reproducible if we keep the material within its linear viscoelastic region. For any one point on the curve, we can determine the stress applied as

$$\sigma = \sigma_0 \sin \omega t \qquad (4.1)$$

where σ is the stress at time t, σ_0 is the maximum stress, ω is the frequency of oscillation, and t is the time. The resulting strain wave shape will depend on how much viscous behavior the sample has as well as how much elastic behavior. In addition, we can write a term for the rate of stress by taking the derivative of the above equation in terms of time:

$$d\sigma / dt = \omega \sigma_0 \cos \omega t \qquad (4.2)$$

We can look at the two extremes of the materials behavior, elastic and viscous, to give us the limiting extremes that will sum to give us the strain wave. Let's start by treating the material as each of the two extremes discussed in Chapter 2. The material at the spring-like or Hookean limit will respond elastically with the oscillating stress. The strain at any time can be written as

$$\varepsilon(t) = E \sigma_0 \sin(\omega t) \qquad (4.3)$$

where $\varepsilon(t)$ is the strain at anytime t, E is the modulus, σ_0 is the maximum stress at the peak of the sine wave, and ω is the frequency. Since in the linear region σ and ε are linearly related by E, we can also write that

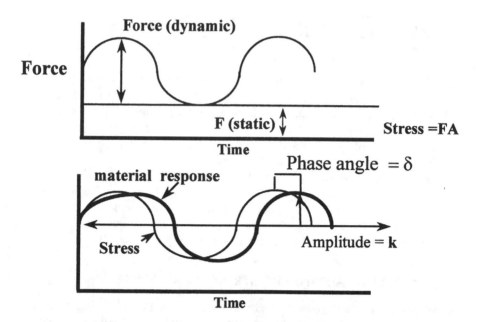

FIGURE 4.1 Oscillating a sample. When a sample is subjected to a sinusoidal oscillating stress, it responses in a similar strain wave provided the material stays within its elastic limits.

$$\varepsilon(t) = \varepsilon_0 \sin(\omega t) \tag{4.4}$$

where ε_0 is the strain at the maximum stress. This curve, shown in Figure 4.2a, has no phase lag (or no time difference from the stress curve), and is called the in-phase portion of the curve.

The viscous limit was expressed as the stress being proportional to the strain rate, which is the first derivative of the strain. So if we take the dashpot we discussed before, we can write a term for the viscous response in terms of strain rate as

$$\varepsilon(t) = \eta d\sigma_0 / dt = \eta \omega \sigma_0 \cos(\omega t) \tag{4.5}$$

or

$$\varepsilon(t) = \eta \omega \sigma_0 \sin(\omega t + \pi/2) \tag{4.6}$$

where the terms are as above and η is the viscosity. We can also replace terms as above to write this as

$$\varepsilon(t) = \omega \varepsilon_0 \cos(\omega t) = \omega \varepsilon_0 \sin(\omega t + \pi/2) \tag{4.7}$$

This curve is shown in Figure 4.2b. Now, let's take the behavior of the material that lies between these two limits. That curve is shown in Figure 4.2c and is intermediate

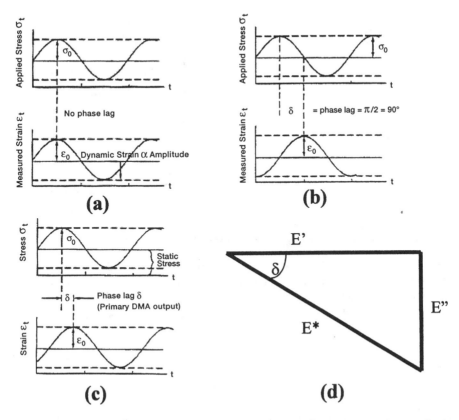

FIGURE 4.2 Responses. When the material responds to the applied stress wave as a perfectly elastic solid, an in-phase response is seen (a), while a purely viscous material gives an out-of-phase response (b). Viscoelastic materials fall in between these two lines, as shown in (c). The relationship between the phase angle, E^*, E', and E'', is graphically shown in (d). (Used with the permission of the Perkin-Elmer Corp., Norwalk, CT.)

between the above cases. The difference between the applied stress and the resultant strain is an angle, δ, and this must be added to equations. So the elastic response at any time can now be written as:

$$\varepsilon(t) = \varepsilon_o \sin(\omega t + \delta) \tag{4.8}$$

From this we can go back to our old trigonometry book and rewrite this as:

$$\varepsilon(t) = \varepsilon_o \left[\sin(\omega t) \cos \delta + \cos(\omega t) \sin \delta \right] \tag{4.9}$$

We can now break this equation, corresponding to the curve in Figure 4.2c, into the in-phase and out-of-phase strains that corresponds to curves like those in Figure 4.2a and 4.2b, respectively. These sum to the curve in 4.2c and are

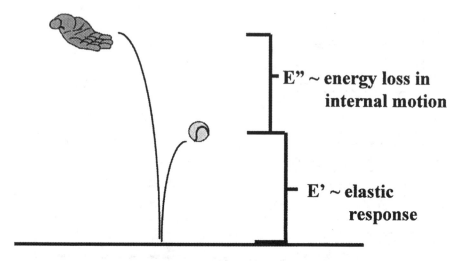

FIGURE 4.3 Storage and loss. When a ball is bounced, the energy initially put in is divided into two parts, a recovered part (how high it bounced) that can be described as E', and the energy lost to friction and internal motions (the difference between the height dropped from and the bounce) called E''.

$$\varepsilon' = \varepsilon_0 \sin(\delta) \qquad\qquad (4.10)$$

$$\varepsilon'' = \varepsilon_0 \cos(\delta) \qquad\qquad (4.11)$$

and the vector sum of these two components gives us the overall or complex strain on the sample

$$\varepsilon^* = \varepsilon' + i\varepsilon'' \qquad\qquad (4.12)$$

This relationship can be seen as the triangle shown in Figure 4.2d and makes sense mathematically.[1] But what does it mean physically in terms of analyzing the polymer behavior?

Basically, this approach allows us to "break" a single modulus (or viscosity or compliance) into two terms, one related to the storage of energy and another related to the loss of energy. This can be seen schematically in the bouncing ball in Figure 4.3. One term lets us see how elastic the polymer is (its spring-like nature), while the other lets us see its viscous behavior (the dashpot). In addition, because all this requires is one sinusoidal oscillation of the polymer, we can get this information quickly rather than through modulus mapping or capillary flow studies.

4.2 CALCULATING VARIOUS DYNAMIC PROPERTIES

Based on our discussion above, a material that under sinusoidal stress has some amount of strain at the peak of the sine wave and an angle defining the lag between the stress sine wave and the strain sine wave. All of the other properties for the DMA are calculated from these data. We can first calculate the storage or elastic modulus, E'. This value is a measure of how elastic the material is and ideally is equivalent to Young's modulus. This is not true in the real world for several reasons. First, Young's modulus is normally calculated over a range of stresses and strains, as it is the slope of a line, while the E' comes from what can be considered a point on the line. Secondly, the tests are very different, as in the stress–strain test, one material is constantly stretched, whereas it is oscillated in the dynamic test. If we were to bounce a ball, as shown in Figure 4.3, the storage modulus (also called the elastic modulus, the in-phase modulus, and the real modulus) could be related to the amount of energy the ball gives back (how high it bounces). E' is calculated as follows:

$$E' = \left(\sigma^{\circ}/\varepsilon^{\circ}\right)\cos\delta = \left(f_{o}/b\mathbf{k}\right)\cos\delta \qquad (4.13)$$

where δ is the phase angle, b is the sample geometry term, f_{o} is the force applied at the peak of the sine wave, and $\mathbf{k}$ is the sample displacement at peak. The full details of developing this equation (and still the best treatment of how DMA equations are derived) are in Ferry.[2]

The amount the ball doesn't recover is the energy lost to friction and internal motions. This is expressed as the loss modulus, E'', also called the viscous or imaginary modulus. It is calculated from the phase lag between the two sine waves as

$$E'' = \left(\sigma^{\circ}/\varepsilon^{\circ}\right)\sin\delta = \left(f_{o}/b\mathbf{k}\right)\sin\delta \qquad (4.14)$$

where δ is the phase angle, b is the sample geometry term, f_{o} is the force applied at the peak of the sine wave, and $\mathbf{k}$ is the sample displacement at the peak.

The tangent of the phase angle is one of the most basic properties measured. Some earlier instruments only recorded phase angle, and consequently the early literature uses the tan d as the measure for many properties. This property is also called the damping, and is an indicator of how efficiently the material loses energy to molecular rearrangements and internal friction. It is also the ratio of the loss to the storage modulus and therefore is independent of geometry effects. It is defined as

$$\tan\delta = E''/E' = \eta'/\eta'' = \varepsilon''/\varepsilon' \qquad (4.15)$$

where η' is the energy loss portion of the viscosity and η'' the storage portion. Because it is independent of geometry (the sample dimensions cancel out above), tan δ can be used as a check on the possibility of measurement errors in a test. For example, if the sample size is changed and the forces are not adjusted to keep the stresses the same, the E' and E'' will be different (because modulus is a function of the stress; see Eq. (4.13) and (4.14) but the tan δ will be unchanged. A change in modulus with no change in the tan δ should lead one to check the applied stresses to see if they are different.

TABLE 4.1

Calculations of Material Properties by DMA

Damping	$\tan \delta = E''/E'$
Complex modulus	$E^* = E' + iE'' = \text{SQRT}(E'^2 + E''^2)$
Complex shear modulus	$G^* = E^*/2(1 + v)$
Complex viscosity	$\eta^* = 3G^*/\omega = \eta'' + i\eta'$
Complex compliance	$J^* = 1/G^*$

Once we have calculated the basic properties all the other properties are calculated from them. Table 4.1 shows the calculation of the remaining properties from DMA. Note that complex viscosity has a dependence on the frequency in the denominator so that at 1 Hz, complex viscosity will overlap the complex modulus. Also note that converting E into G or the reverse requires the use of Poisson's ratio, v. Out of these remaining properties, the most commonly used is the complex viscosity, η^*. The reasons for this are discussed in Chapter 7. The biggest reason is that data from frequency scans give a viscosity vs. shear rate (or frequency) curve that can be obtained much faster than by other methods. The complex viscosity is given by

$$\eta^* = G^*/\omega = E^*/2(1 + v) \tag{4.16}$$

where G^* is the complex shear modulus. Like other complex properties, it can be divided into an in-phase and out-of-phase component:

$$\eta^* = \eta' - i\eta'' \tag{4.17}$$

where η' is a measure of energy loss and η'' is a measure of stored energy. Unlike the difficulties that sometimes exist with E' and Young's modulus, the complex viscosity usually agrees fairly well with the steady shear viscosity. One applies the Gleissele's mirror relationship, which is

$$\eta(d\gamma/dt) = \eta^*(t)\big|_{t=1/(d\gamma/dt)} \tag{4.18}$$

where $\eta^*(t)|$ is the limiting value of the viscosity as $d\gamma/dt$ approaches 0.[3] Another option, the well-known Cox–Merz rule, is discussed in Chapter 7. The agreement of both is normally within ±10%. The relationship between steady shear viscosity, normal forces, and dynamic viscosity is discussed in more detail in Chapter 7.

4.3 INSTRUMENTATION FOR DMA TESTS

4.3.1 FORCED RESONANCE ANALYZERS

The most common analyzers on the market today are forced resonance analyzers. These are designed to force the sample to oscillate at a fixed frequency and are ideally suited for scanning material performance across a temperature range. The analyzers consist of several parts for controlling the deformation, temperature, sample geometry, and sample environment. Figure 4.4 shows an example. Obviously, various choices can be made for all of the components, each with their own advantages and disadvantages. For example, furnaces come in a wide variety of materials: the analyzer shown has a ceramic-covered, wire-wound type which is used because it gives very fast temperature response over a wide range and is very inert. However, it is more fragile than other types. Other choices could be a Peltier heater or a forced-air furnace. The most important issue is accurate and reproducible temperature control, as discussed in Chapter 5. Similarly, the measurement of the probe position can be done by using several techniques including a linear vertical displacement transducer, an optical encoder, or an eddy current detector.

4.3.2 STRESS AND STRAIN CONTROL

One of the biggest choices made in selecting a DMA is to decide whether to chose stress (force) or strain (displacement) control for applying the deforming load to the sample (Figure 4.5a and b). Strain-controlled analyzers, whether for simple static testing or for DMA, move the probe a set distance and use a force balance transducer or load cell to measure the stress. These parts are typically located on different shafts. The simplest version of this is a screw-driven tester, where the sample is pulled one turn. This requires very large motors, so the available force always exceeds what is needed. They normally have better short time response for low viscosity materials and can normally perform stress relaxation experiments easily. They also usually can measure normal forces if also torsional analyzers. A major disadvantage is that their transducers may drift at long times or with low signals.

Stress-controlled analyzers are cheaper to make because there is only one shaft, but somewhat trickier to use. Many of the difficulties have been alleviated by software, and many strain-controlled analyzers on the market are really stress-controlled instruments with feedback loops making them act as if they were strain-controlled. In stress control, a set force is applied to the sample. As temperature, time, or frequency varies, the applied force remains the same. This may or may not be the same stress: in extension, for example, the stretching and necking of a sample will change the applied stress seen during the run. However, this constant stress is a more natural situation in many cases, and it may be more sensitive to material changes. Good low force control means they are less likely to destroy any structure in the sample. Long relaxation times or long creep studies are more easily performed on these instruments. Their biggest disadvantage is that their short time responses are limited by inertia with low viscosity samples.

Since most DMA experiments are run at very low strains (~0.5% maximum) to stay well within a polymers linear region, it has been reported that the both analyzers give the same results. However, when one gets to the nonlinear region, the difference becomes significant, as stress and strain are no longer linearly related. Stress control can be said to duplicate real-life conditions more accurately, since most applications of polymers involve resisting a load.

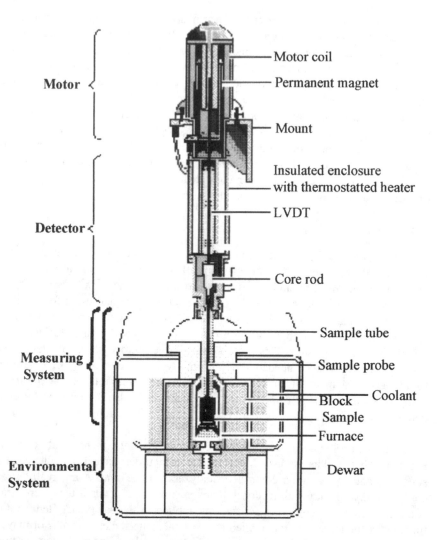

FIGURE 4.4 **Schematic of the Perkin-Elmer DMA-7e** showing the motor, LVDT, sample compartment, furnace, and heat sink. (Used with the permission of the Perkin-Elmer Corp., Norwalk, CT.)

4.3.3 AXIAL AND TORSIONAL DEFORMATION

DMA analyzers are normally built to apply the stress or strain in two ways (Figure 4.5c and 4.5d). One can apply force in a twisting motion so one is testing the sample in torsion. This type of instrument is the dynamic analog of the constant shear spinning disk rheometers. While mainly used for liquids and melts, solid samples may also tested by twisting a bar of the material. Torsional analyzers normally also permit continuous shear and normal force measurements. Most of these analyzers can also do creep–recovery, stress–relaxation, and stress–strain experiments.

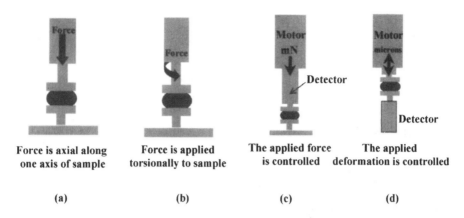

Force is axial along   Force is applied   The applied force   The applied
one axis of sample   torsionally to sample   is controlled   deformation is controlled

(a)   (b)   (c)   (d)

FIGURE 4.5   Types of DMAs: (a) axial, (b) torsional, (c) controlled stress, and (d) controlled strain.

Axial analyzers are normally designed for solid and semisolid materials and apply a linear force to the sample. These analyzers are usually associated with flexure, tensile, and compression testing, but they can be adapted to do shear and liquid specimens by proper choice of fixtures. Sometimes the instrument's design makes this inadvisable, however. (For example, working with a very fluid material in a system where the motor is underneath the sample has the potential for damage to the instrument if the sample spills into the motor.) These analyzers can normally test higher modulus materials than torsional analyzers and can run TMA studies in addition to creep–recovery, stress–relaxation, and stress–strain experiments.

Despite the traditional selection of torsional instruments for melts and liquids and axial instruments for solids, there is really considerable overlap between the types of instruments. With the proper choice of sample geometry and good fixtures, both types can handle similar samples, as shown by the use of both types to study the curing of neat resins (see the data in Chapter 6 for examples of this). Normally axial analyzers can not handle fluid samples below about 500 Pa-seconds and torsional instruments will top out with the harder samples (the exact modulus depending on the size of the motor and/or load cell.) Because of this considerable overlap, I will not be separating the applications by analyzer type in the following discussions.

4.3.4   FREE RESONANCE ANALYZERS

Some of the earlier instruments used the approach used in free resonance analyzers, but I am discussing it last because it requires a little more sophistication than the conceptually simpler forced resonance analyzers. If we suspend a sample in it that can swing freely, it will oscillate like a harp or guitar string until the oscillations gradually come to a stop. The naturally occurring damping of the material controls the decay of the oscillations. This gives us a wave, shown in Figure 4.6a, which is

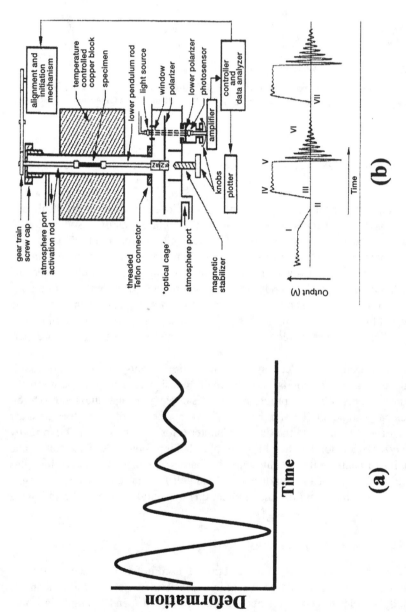

FIGURE 4.6 Free resonance analyzer: (a) the decay wave from free resonance and (b) a schematic of a Torsional Braid Analyzer. (From John K. Gillham and John B. Enns, "On the cure and properties of thermosetting polymers using Torsional Braid Analysis," *Trends in Polymer Science*, 2, 12, 1994, pp. 406–419. With permission from Elsevier Science.)

a series of sine waves decreasing in amplitude and frequency. Several methods exist to analyze these waves and are covered in the review by Gillham.[4] These methods have also been successfully applied to the recovery portion of a creep–recovery curve where the sample goes into free resonance on removal of the creep force.[5]

From the decay curve, we need to calculate or collect the period, T, and the logarithmic decrement, Λ. Several methods exist for both manual and digital processing.[4,6] Fuller details of the following may be found in McCrum et al.[6] and Gillham.[4] Basically, we begin by looking at the decay of the amplitude over as many swings as possible to reduce error:

$$\Lambda = 1/j \, \ln\left(A_n/A_{(n+j)}\right) \tag{4.19}$$

where j is the number of swings and A_n is the amplitude of the nth swing. For one swing, where $j = 1$, the equation becomes

$$\Lambda = \ln\left(A_n/A_{(n+j)}\right) \tag{4.20}$$

if for a low value of Λ where A_n/A_{n+1} is approximately 1, we can rewrite the equation as

$$\Lambda \approx \tfrac{1}{2}\left(\left(A_n^2 - A_{n+1}^2\right)/A_n^2\right) \tag{4.21}$$

From this, since the square of the amplitude is proportional to the stored energy, $\Delta W/W_{st}$, and the stored energy can be expressed as $2\pi \tan \delta$, the equation becomes

$$\Lambda \approx \tfrac{1}{2}\left(\Delta W/W_{st}\right) = \pi \tan \delta \tag{4.22}$$

which gives us the phase angle, δ. The time of the oscillations, the period, T, can be found using the following equation:

$$T = 2\pi \sqrt{\frac{M}{\Gamma_1}} \sqrt{\frac{1+\Lambda^2}{4\pi^2}} \tag{4.23}$$

where Γ_1 is the torque for one cycle and M is the moment of inertia around the central axis. Alternatively, we could calculate the T directly from the plotted decay curve as

$$T = (2/n)\left(t_n - t_0\right) \tag{4.24}$$

where n is the number of cycles and t is time. From this, we can now calculate the shear modulus, G, which for a rod of length, L, and radius, r, is

$$G = \left(\frac{4\pi^2 ML}{NT^2}\right)\left(1 + \frac{\Lambda^2}{4\pi^2}\right) - \left(\frac{mgr}{12N}\right) \tag{4.25}$$

where m is the mass of the sample, g the gravitational constant, and N is a geometric factor. In the same system, the storage modulus, G', can be calculated as

$$G' = \left(1/T^2\right)\left(8\pi^2 ML/r^4\right) \tag{4.26}$$

where I is the moment of inertia for the system. Having the storage modulus and the tangent of the phase angle, we can now calculate the remaining dynamic properties.

Free resonance analyzers normally are limited to rod or rectangular samples or materials that can be impregnated onto a braid. This last approach is how the curing studies on epoxy and other resin systems were done in torsion and gives these instruments the name of Torsional Braid Analyzers (TBA). A schematic of a TBA is shown in Figure 4.6b.

4.4 FIXTURES OR TESTING GEOMETRIES

We are now going to take a quick look at the commonly used fixtures for the DMA. It is important to be familiar with the effect the sample geometry has on stress and strain values, because small dimensional changes often have large consequences. For measuring the temperature of transitions alone, a great deal of inaccuracy can be tolerated in the sample dimensions. This is not true for modulus values. Each of the geometries discussed below has a different set of equations for calculating stress and strain from force and deformation. One way of handling this is the use of a geometry factor, b, used to calculate E' and E'' in Eq. (4.13) and (4.14). The equations for these geometric factors are given in Table 4.2. One practical use of these equations is in estimating the amount of error in the modulus expected due to the inaccuracies of measuring the sample. For example, if the accuracy of the dimensional measurement is 5%, the error in the applied stress can be calculated. Since many of these factors involve squared and cubed terms, small errors in dimensions can generate surprisingly large errors in the modulus.

Two other issues need to be mentioned: those of inertia effects and of shear heating. Inertia is the tendency of an object to stay in its current state, whether moving or at rest. When a DMA probe is oscillated, this must be overcome and as the frequency increases, the effect of the instrument inertia becomes more troublesome. Shear heating occurs when the mechanical energy supplied to the sample is converted to heat by the friction and changes the sample temperature. Both of these can occur in either analyzer, but they are more common in fluid samples.

TABLE 4.2
Geometric Factors for Fixtures

3 pt. bending bar	$x_s^3/4z_sy_s^3$	Parallel Plate in Shear	$y_s/2\pi r_s^2$
rod	$4x_s^3/6\pi r_s^4$	Cone and Plate	$\Theta/2\pi r_s^3$
Single Cantilever	$4x_s^3/z_sy_s^3$	Couette	$\dfrac{1-(R_1/R_2)^2}{2\pi L(R_1)^2}$
Dual Cantilever	$x_s^3/16z_sy_s^3$	Torsion Bar	$\dfrac{3+8(z_s/x_s)^2)}{x_sz_s^3/y_s(1-0.378)(z_s/x_s)^2))}$
Extension	y_s/x_sz_s		
Plates in Compression	$y_s/2\pi r_s^2$		
Shear Sandwich	$(y_s/2\pi r_s^2)$		

Note: These factors are used to convert the load and the amount of deformation into stress and strain by using the dimensions of the sample. The letters refer to dimensions as drawn in Figures 4.7–4.12. The dimensions for rectangular samples are width (x), height (y), and depth (z), as drawn in the respective fixture. For sample with a circular cross-section (i.e., a disk, plate, bar, or fiber), r_s is the sample radius and y is again the height. Θ is the cone angle in a cone-and-plate fixture. For the Couette fixture, R_1 and R_2 refer to the inner and outer cylinder diameters, respectively, while L refers to the length of the inner cylinder.

Source: Used with the permission of the Perkin-Elmer Corp., Norwalk, CT and of Rheometric Scientific, Piscataway, NJ.

4.4.1 AXIAL

Axial analyzers allow a great deal of flexibility in the choice of fixtures, which allows for the testing of a wide range of materials. Ease of exporting data is also important in these analyzers, as they are often adapted to run very specialized tests. For example, some contact lens manufacturers test samples on spherical plates and tubing manufacturers will use fixtures that test the tubing in its original shape. With easy access to the raw data, one can go to Roark's[7] and look up the stress and strain formulas for the appropriate geometry.

4.4.1.1 Three-Point and Four-Point Bending

There are four types of bending or flexure fixtures commonly used. The simplest and most straightforward of these is the three-point bending fixture shown in Figure 4.7a. No flexure mode is a pure deformation, as they can all be looked at as a combination of an extension and compressive strain. Three-point bending depends on the specimen being a freely moving beam, and the sample should be about 10%

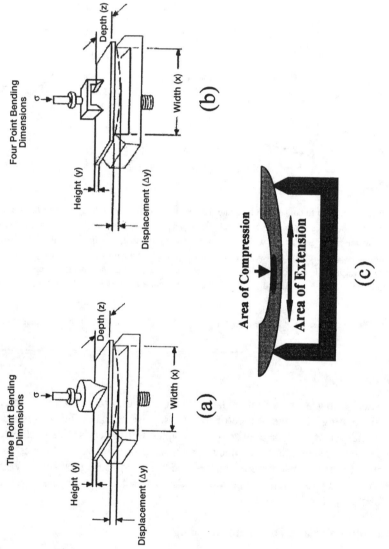

FIGURE 4.7 Flexure test fixtures I: (a) three-point bending, (b) four-point bending, and (c) compressive and tensile strains in a three-point bending specimen. (Used with the permission of the Perkin-Elmer Corp., Norwalk, CT.)

longer on each end than the span. The four sides of the span should be true, i.e., parallel to the opposite side and perpendicular to the neighboring sides. There should be no nicks or narrow parts. Rods should be of uniform diameter. Throughout the experiment the beam should be freely pivoting: this is checked after the run by examining the sample to see if there are any indentations in the specimen. If there are, this suggests that a restrained beam has been tested, which gives a higher apparent modulus. The sample is loaded so the three edges of the bending fixture are perpendicular to the long axis of the sample.

Four-point bending replaces the single edge on top with a pair of edges, as shown in Figure 4.7b. The applied stress is spread over a greater area than in three-point bending, as the load is applied on two points rather than one. The mathematics remains the same, as the flexing of the beam is unchanged. One can use either knife-edged or round-edged fixtures and, in some cases, a ball probe is used to replace the center knife-edge of the three-point bending fixture to simplify alignment. As long as the specimen shows the required even deflection, these are all acceptable. Sample alignment and deformation are shown in Figure 4.7c.

4.4.1.2   Dual and Single Cantilever

Cantilever fixtures clamp the ends of the specimen in place, introducing a shearing component to the distortion (Figure 4.8a) and increasing the stress required for a set displacement. Two types of cantilever fixtures are used: dual cantilever, shown in Figure 4.8b, and single cantilever, shown in Figure 4.8c. Both cantilever geometries require the specimen to be true as described above and to be loaded with the clamps perpendicular to the long axis of the sample. In addition, care must be taken to clamp the specimen evenly, with similar forces, and not to introduce a twisting or distortion in clamping. Moduli from dual cantilever fixtures tend to run 10–20% higher than the same material measured in three–point bending. This is due to shearing strain induced by clamping the specimen in place at the ends and center, which makes the sample more difficult to deform.

4.4.1.3   Parallel Plate and Variants

Parallel plate in axial rheometers means testing in compression, and several variations of simple parallel plates exist for special cases (Figure 4.9a, 4.9b, and 4.9c). Sintered or sandblasted plates are used for slippery samples, plates and trays for samples that drip or need to be in contact with a solvent, and plates and cups for low viscosity materials. Photocuring materials can be studied with quartz fixtures. For samples in compression, circular plates are normally used, because these are easily manufactured and the samples can be fabricated by die-cutting films or sheets to size. Rectangular plates and samples are also used. Any type of plate needs to be checked after installation to make sure the plates are parallel. The easiest way to check the alignment is by bringing the plates together and seeing if they seeing if they are touching each other with no spaces or gaps. Samples should be the same size as the plates with the edges even or flush, having no dips or bulges. On compressing, they should deform by bowing outward slightly in a even, uniform bulge.

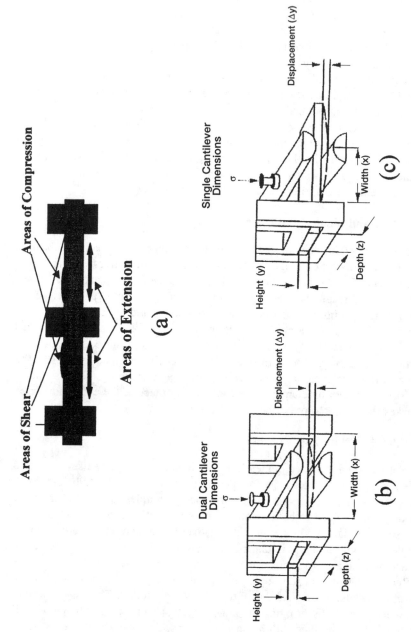

FIGURE 4.8 Flexure test fixtures II: (a) strain in a dual cantilever specimen showing the shear regions, (b) dual cantilever, and (c) single cantilever. (Used with the permission of the Perkin-Elmer Corp., Norwalk, CT.)

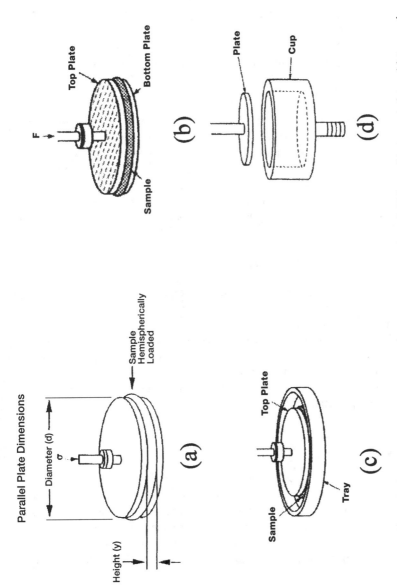

FIGURE 4.9 Parallel plate fixture for axial analyzers: (a) standard parallel plates, (b) sintered plates, (c) tray and plates, and (d) cup and plate. The latter is also used for bulk measurements. (Used with the permission of the Perkin-Elmer Corp., Norwalk, CT.)

4.4.1.4 Bulk

If we run a sample in a dilatometer-like fixture (Figure 4.9d) where it is restrained on all sides, we measure the bulk modulus of the material. This can be done in a specialized Pressure Volume Temperature (PVT) instrument with very high loads (up to 200 MPa of applied pressure) to study the free volume of the material.[8] In this geometry, alignment is critical, as the fit between the top plate and the walls of the cup must be very tight to prevent the material from escaping. The cup and plate/plunger system must still be able to move freely without binding.

4.4.1.5 Extension/Tensile

Extension or tensile analysis (Figure 4.10a) is done on samples of all types and is one of the more commonly done experiments. This geometry is more sensitive to loading and positioning of the sample than most other geometries. Any damage or distress to the edges of the sample as well will cause inaccuracies in the measurements. A nick in the edge will also often cause early failure, as it acts as a stress concentrator. After loading a film or fiber in extension, it is important to adjust it so that there are not any twists, the sides are perpendicular to the bottom, and there are no crinkles.

4.4.1.6 Shear Plates and Sandwiches

Figure 4.10b shows one of the two approaches to measuring shear in an axial analyzer. Shear measurements are done by a sliding plate moving between two samples. This is the common approach today. Another older approach is to use a single sample, but this requires a very rigid analyzer. Again, sample size and shape must be controlled tightly for modulus data to be fully meaningful. It is very important in the shear sandwich fixtures to make sure both samples are as close to identical as possible. This technique can be difficult to run over wide temperature ranges, as the thermal expansion of the fixture can cause the clamping force to vary greatly.

4.4.2 TORSIONAL

Samples run in torsional analyzers tend to be of lower viscosity and modulus than those run in axial instruments. Torsional instruments can be made to handle a wide variety of materials ranging from very low viscosity liquids to bars of composite. Inertia affects tend to be more of a problem with these instruments, and very sophisticated approaches have been developed to address them.[9] Most torsional rheometers can also perform continuous shear experiments and also measure normal forces.

4.4.2.1 Parallel Plates

The simplest geometry in torsional shear is two parallel plates run at a set gap height. This is shown in Figure 4.11a. Note that the important dimensions are the same as

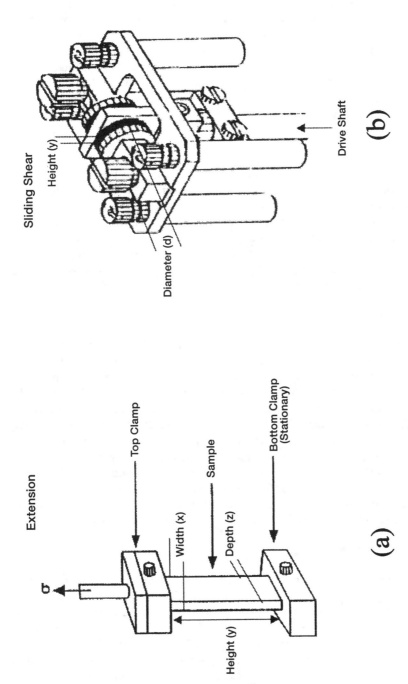

FIGURE 4.10 Extension and sliding shear fixtures: (a) an extension fixture for thin films or fibers. (Used with the permission of the Perkin-Elmer Corp., Norwalk, CT.) (b) A sliding plate shear fixture using two round samples. (Used with the permission of Rheometric Scientific, Piscataway, NJ.)

in Figure 4.10a. The height or gap here is determined by the viscosity of the sample. We want enough space between the plates to obtain decent flow behavior, but not so much that the material flies out of the instrument. The edge of the sample should be spherical without fraying or rippling. These plates have an uneven strain field across them: the material at the center of the plate is strained very little, as it barely moves. At the edge, the same degree of turning corresponds to a much larger movement. So the measured strain is an average value and the real strain is inhomogeneous. The thrust against the plates can be used to calculate the normal stress difference in steady shear runs.

4.4.2.2 Cone-and-Plate

The cone-and-plate geometry uses a cone of known angle instead of a top plate. When this cone angle is very small, the system generates an even, homogeneous strain across the sample. Shown in Figure 4.11b, the gap is set to a specific value, normally supplied by the manufacturer. This value corresponds to the truncation of the cone. The cone-and-plate system is probably the most common geometry used today for studying non-Newtonian fluids. As above, at very high shear rate, the material reaches a critical edge velocity and fails. This geometry is discussed extensively by Macosko along with other geometries for testing fluids.[10]

4.4.2.3 Couette

Some materials are of such a low viscosity that testing them in a cone-and-plate or parallel plate fixture is inadvisable. When this occurs, one of the Couette geometries can be used. Also called concentric or coaxial cylinders, the geometry is shown in Figure 4.12a with the recessed bottom style of bob (inner cylinder). This shape is used to eliminate or reduce end effects. Other choices might be a conicylinder, where the cylinder has a cone-shaped (pointed) end or a double Couette, where a thin sheet of material has solution around it. The recessed end traps air, which transfers no force to the fluid, and this seems to be the one most commonly used today. This fixture requires tight tolerances, and the side gap is fixed by the design. The bottom gap is also tightly controlled and the tops of the cylinders must be flush. The fluid level must come close, say within 5 mm of the top of the fixture, but not overflow it. Both inertia effects and shear heating are concerns that must be addressed.

4.4.2.4 Torsional Beam and Braid

Stiff, solid samples in a torsional analyzer are tested as bars or rods that are twisted about their long axis. This geometry is shown in Figure 4.12b. The bar needs to be prepared as precisely as those discussed above. Another variation is the use of a braid of material impregnated with resin for curing studies. This can be a tricky approach as even nondrying oils appear to increase in viscosity or cure (cross-linked) as they heat up (their viscosity drops and the fibers start rubbing together so that the measured viscosity appears to increase as if the material had cured).

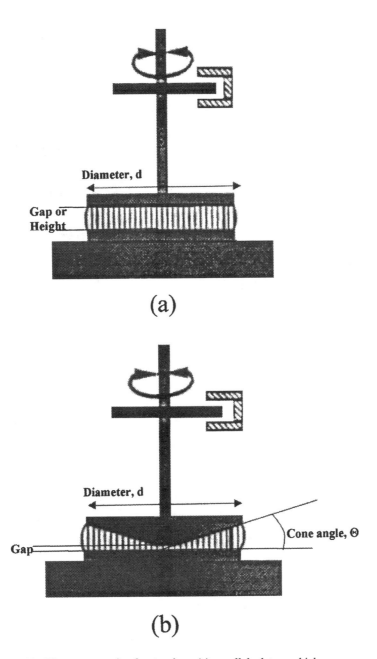

FIGURE 4.11 Plate geometries for torsion: (a) parallel plates, which are measured the same way as in Figure 4.9, and (b) cone and plates. (Used with the permission of Rheometric Scientific, Piscataway, NJ.)

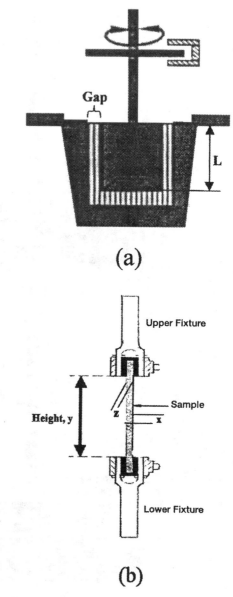

FIGURE 4.12 Couette and torsion bar geometries: (a) Couette fixture and (b) torsion bar, showing a rod-shaped specimen. (Used with the permission of Rheometric Scientific, Piscataway, NJ.)

4.5 CALIBRATION ISSUES

Calibration may be one of the most misunderstood concerns in operating a DMA. All of the systems in a DMA need to be tied back to some standard for the collected data to have meaning. Figure 4.13 shows the Perkin-Elmer DMA, as an example, with the calibrations needed to standardize the performance of each part of the instrument. All instruments require this, and normally these procedures are described

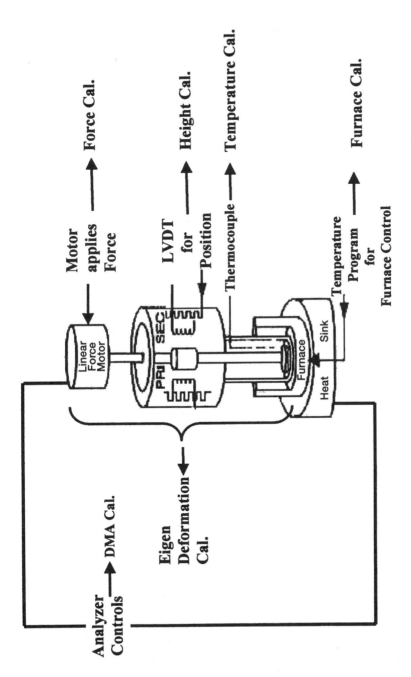

FIGURE 4.13 Calibration. A schematic showing the interrelationship of calibration and analyzer systems. (Used with the permission of the Perkin-Elmer Corp., Norwalk, CT.)

in the operation manual. Temperature calibration is especially critical, as material properties are strongly influenced by it.

The calibrations shown in Figure 4.13 cover the major types used. Most analyzers require some sort of system or self-check that verifies all of the components are working or installed. Most instruments offer a self-diagnostic test for this purpose. This is normally run either before calibration or after a specific calibration fails. Force, height or position, and temperature calibrations relate the instrument's signal to a known standard of performance. This should ideally be traceable to a primary source. For temperature, National Institute for Standards and Technology (NIST) traceable materials, such as indium or tin, are preferred.

All instruments have some sort of movement, as none are ideally rigid. Therefore, some measure of the instrument's rigidity or self-deformation is needed. An eigen-deformation test is one way to do this. Various approaches are used, but as harder and stiffer samples are run, the amount of deformation in the instrument becomes more important. With very stiff samples, the deformation in the analyzer could become greater than the sample and inaccurate results are obtained. Similarly, if the analyzer is too stiff, then very soft samples may not be detected.

Finally, some sort of control over the furnace's performance is needed. Whether a control thermocouple or a voltage table, the furnace need to be tuned so that its operation is in agreement with the measured temperature of the thermocouple. This is normally done by a procedure in which the instrument maintains an isothermal hold at a set furnace voltage and adjusts the voltage table to the correct thermocouple. In addition, the approach of the furnace to the set temperature should be tuned to give the desired response (i.e., degree of overshoot). This is done by using proportional, integral, and derivative control (PID) files in many cases. This approach has the advantage of a great detail of control in tuning the furnace.

4.6 DYNAMIC EXPERIMENTS

DMA experiments can be classed as temperature–time studies, frequency studies, and dynamic stress–strain curves. Temperature–time scans hold the frequency constant as the temperature or time at temperature changes. These are discussed in Chapters 5 and 6. Frequency scans vary the frequency at a set temperature and are discussed in Chapter 7. Some techniques exist that combine these two into one experiment, and these are mentioned in Chapter 7. Finally the dynamic force can be constantly increased at a fixed rate and a dynamic stress–strain curve can be generated.

This last technique is commonly used as a method of tuning the DMA and selecting the proper operating range for a sample. It is done by increasing the amplitude of the sine wave analogously to the increased stress in a stress–strain experiment. This is shown in Figure 4.14. It is often necessary to also continually increase the static force to maintain constant tension on the sample. Because this experiment is a series of larger and larger sine waves, we can not only plot dynamic stress vs. dynamic strain, but also get E', E'', E*, tan δ, η*, and other dynamic properties for each cycle. This allows us to plot those properties as a function of dynamic stress too. Besides providing insight into the material behavior, this method

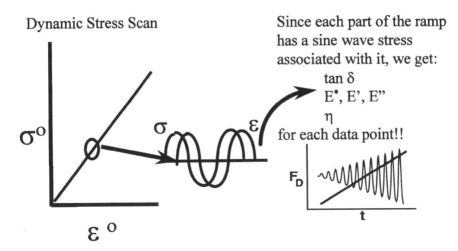

Dynamic Stress Scan

Since each part of the ramp has a sine wave stress associated with it, we get:

tan δ
E*, E', E"
η
for each data point!!

σ^o

σ

ε

ε^o

F_D

t

FIGURE 4.14 **Dynamic stress–strain curves** involve the application of a dynamic stress ramp to a specimen so each data point measures a different sine wave.

makes an excellent method of quality control. As stated above, each sine wave can be solved for the full range of dynamic properties, allowing very detailed comparisons in very short times. Figure 4.15 shows dynamic stress–dynamic strain and tan δ–dynamic strain plots on two rubbers.

APPENDIX 4.1 CALIBRATION AND VERIFICATION OF AN INSTRUMENT

The calibration of an instrument is normally done in a time period between once a month to a few times a year. It is often desirable to check the calibration of the instrument without having to redo it. The following is the procedure I use to check the calibration of the Perkin-Elmer DMA-7e. Similar approaches can be used with other instruments, such as RheoSci SR-5 or Mark 4 DMTA.

The first thing I check is the position measurement of the instrument. This is simply done by zeroing the probe on the platform and letting it sit for 1–5 minutes. One looks to see if a stable zero is obtained. Then one repeats the test using a sample of known height. The height measurement should be correct and no drift should be seen. See Figure 4.16a.

The accuracy of the applied force is now checked. This can be done by attaching a weight to the tarred probe and running a stress scan. For the DMA-7e, this is done by loading the weight tray and tarring the probe with the tray in place. On the SR-5 or Aries, a special fixture is attached to the instrument to transfer the torsional force to a weight. We then run a stress scan so that the weight moves. In the DMA-7e, we run from −400 to −600 mN for a 50 g weight. When the weight is exceeded, the probe position changes (Figure 4.16b).

The self-deformation of the analyzer can be checked in an axial analyzer by applying a low level of stress to an empty platform and then applying the maximum

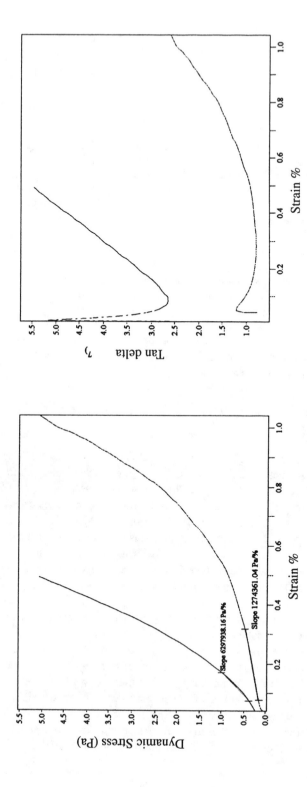

FIGURE 4.15 Results from dynamic stress–strain curves showing (a) the relationship between dynamic stress and strain and (b) tan δ vs. dynamic strain for two materials.

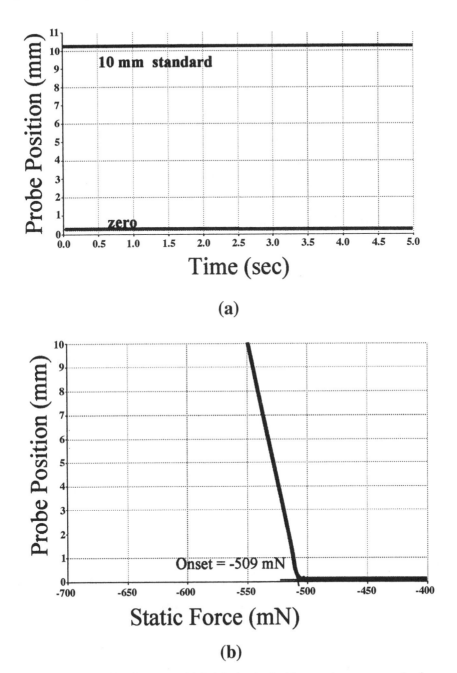

FIGURE 4.16 Verification tests: (a) height is checked by running a zero and a known standard for several minutes, then looking at the difference between the lines as well as any change over time, (b) a stress scan from –400 to –600 mN shows where the 50 mg weight is lifted off the platform. The overlaying of programmed and actual temperature (c) for a run shows how well the furnace is tuned, while, at the same time, probe position (d) is used to check the melting point of Indium.

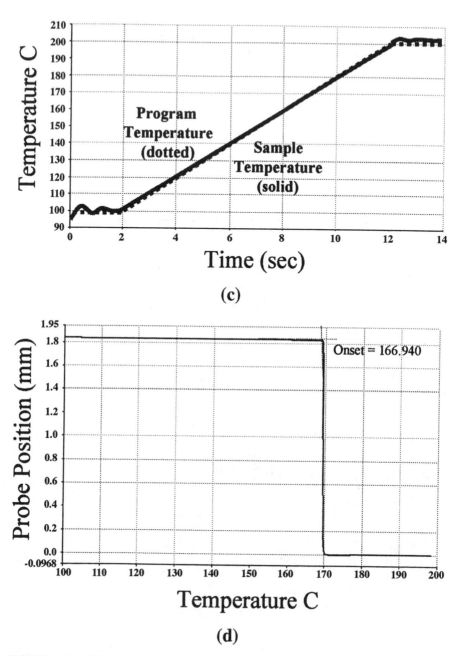

(c)

(d)

FIGURE 4.16 *(Continued).*

possible stress. The amount of change seen in the probe position and amplitude is recorded and compared to the manufacturer's specification. In cases where the measured value is high, a loose fitting or fixture is the most common cause for excessive movement in an analyzer.

Finally, one needs to check both the temperature accuracy and the furnace control. I do this in one run by programming a cycle with 2-minute isothermal holds at 100°C and 200°C and a 10°C per minute ramp between the holds. A sample of indium is loaded under the probe and the analyzer stabilized at 95°C. After the run is complete, one can check both temperature performance (Figure 4.16c) and temperature accuracy (Figure 4.16d).

NOTES

1. This explanation was developed in terms of stress control and tensile stress, instead of the shear strain and strain control normally used. For more detailed development, see one of the following: L. Sperling, *Introduction to Physical Polymer Science,* 2nd ed., Wiley, New York, 1996. L. Nielsen et al., *Mechanical Properties of Polymers and Composites,* Marcel-Dekker, New York, 1996. J. Ferry, *Viscoelastic Properties of Polymers,* Wiley, New York, 1980. The nice thing is the software does all this today and you just need to understand how it works so you realize exactly what you are doing.
2. J. Ferry, *Viscoelastic Properties of Polymers,* Wiley, New York, 1980, Ch. 5–7.
3. R. Armstrong, B. Bird, and O. Hassager, *Dynamics of Polymer Fluids,* vol. 1, *Fluid Mechanics,* 2nd ed., Wiley, New York, 1987, pp.151–153.
4. J. Gillham in *Developments in Polymer Characterizations,* vol. 3, J. Dworkins, Ed., Applied Science Publisher, Princeton, 1982, pp. 159–227. J. Gillham and J. Enns, *TRIP,* 2(12), 406, 1994.
5. U. Zolzer and H-F. Eicke, *Rheologia Acta,* 32, 104, 1993.
6. N. McCrum et al., *Anelastic and Dielectric Properties of Polymeric Solids,* Dover, New York, 1992, pp.192–200.
7. W. Young, *Roark's Formulas for Stress and Strain,* McGraw-Hill, New York, 1989.
8. P. Zoller and Y. Fakhreddine, *Thermochim. Acta,* 238, 397, 1994. P. Zoller and D. Walsh, *Standard Pressure-Volume-Temperature Data for Polymers,* Technomic Publishing, Lancaster, PA 1995.
9. See for example the instrument manuals written by Rheometric Sciences of Piscataway, NJ, where inertia affects are discussed at length.
10. C. Macosko, *Rheology,* VCH Publishers, New York, 1994, Ch. 5.

5 Time–Temperature Scans: Transitions in Polymers

One of most common uses of the DMA for users from a thermal analysis background is the measurement of the various transitions in a polymer. A lot of users exploit the greater sensitivity of the DMA to measure T_g's undetectable by the differential scanning calorimeter (DSC) or the differential thermal analyzer (DTA). For more sophisticated users, DMA temperature scanning techniques let you investigate the relaxation processes of a polymer. In this chapter, we will look at how time and temperature can be used to study the properties of polymers. We will address curing studies separately in Chapter 6.

5.1 TIME AND TEMPERATURE SCANNING IN THE DMA

If we start with a polymer at very low temperature and oscillate it at a set frequency while increasing the temperature, we are performing a temperature scan (Figure 5.1a). This is what most thermal analysts think of as a DMA run. Similarly, we could also hold the material at a set temperature and see how its properties change over time (Figure 5.1b).

Experimentally we need to be concerned with the temperature accuracy and the thermal control of the system, as shown in Figure 5.2. This is one of the most commonly overlooked areas experimentally, as poor temperature control is often accepted to maintain large sample size. A large sample means that there will be a temperature difference across the specimen, which can result in anomalies such as dual glass transitions in a homopolymer.[1] In his Polymer Fluids Short Course,[2] Bird describes experiments where measuring the temperature at various points in a large parallel plate experiment shows a 15°C difference from the plate edge to the center. Large samples require very slow heating rates and hide local differences. This is especially true in post-cure studies. A smaller sample permits a smaller furnace, which is inherently more controllable. Also, smaller sample size allows the sectioning of specimens to see how properties vary across a specimen.

It is often very difficult to examine one specimen across the whole range of interest with only one experiment or one geometry. Materials are very stiff and brittle at low temperatures and soft near the melt, so very different conditions and fixtures may be required. Some analyzers use sophisticated control loops[3] to address this problem, but often it is best handled doing multiple runs.

5.2 TRANSITIONS IN POLYMERS: OVERVIEW

The thermal transitions in polymers can be described in terms of either free volume changes[4] or relaxation times.[5] While the latter tends to be preferred by engineers and

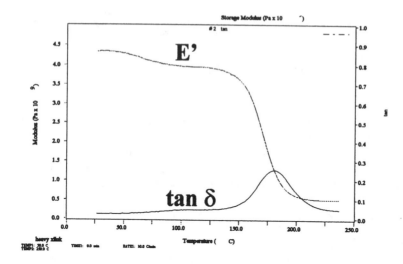

(a) Temperature Scan

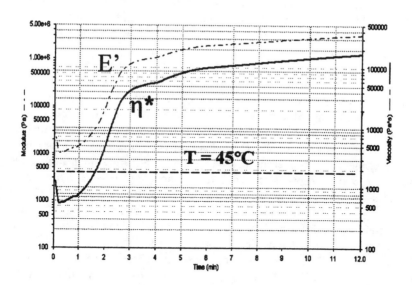

(b) Time Scan

FIGURE 5.1 Time–temperature studies in the DMA: we can (a) vary temperature at a set rate and scan across the various transitions and regions of a material's behavior or (b) hold temperature constant and watch properties change as a function of time, gas changes, etc. In this case, an epoxy cures at 45°C.

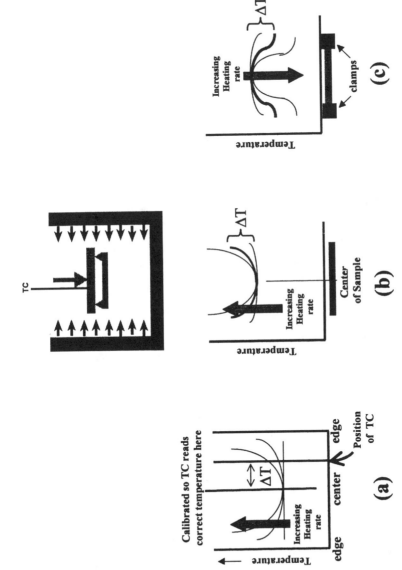

FIGURE 5.2 Temperature control in a DMA sample: A schematic representation of the problems of temperature control in a DMA for a solid sample. Similar problems exist for melts. (a) The problem of thermal lag between the sample and the furnace, which is normally handled by calibration, (b) the lag across a sample under heating due to the thermal conductivity of the sample, (c) the effect of heavy clamps on sample temperature.

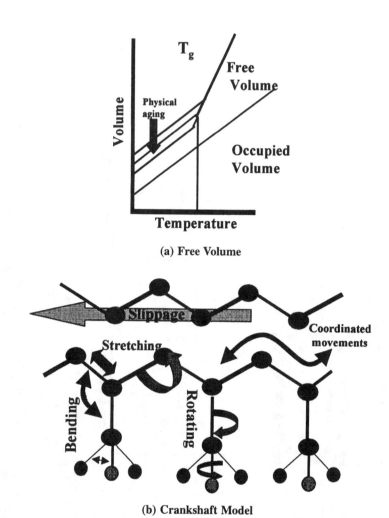

(a) Free Volume

(b) Crankshaft Model

FIGURE 5.3 Free volume, v^f, in polymers: (a) the relationship of free volume to transitions, and (b) a schematic example of free volume and the crankshaft model. Below the T_g in (a) various paths with different free volumes exist depending on heat history and processing of the polymer, where the path with the least free volume is the most relaxed. The crankshaft model (b) shows the various motions of a polymer chain. Unless enough free volume exists, the motions cannot occur.

rheologists in contrast to chemist and polymer physicists who lean toward the former, both descriptions are equivalent. Changes in free volume, v^f, can be monitored as a volumetric change in the polymer; by the absorption or release of heat associated with that change; the loss of stiffness; increased flow; or by a change in relaxation time.

The free volume of a polymer, v^f, is known to be related to viscoelasticity,[6] aging,[7] penetration by solvents,[8] and impact properties.[9] Defined as the space a molecule has for internal movement, it is schematically shown in Figure 5.3a. A simple approach to looking at free volume is the crankshaft mechanism,[10] where the molecule is imagined as a series of jointed segments. From this model, we can simply describe the various transitions seen in a polymer. Other models exist that allow for more precision in describing behavior; the best seems to be the Doi–Edwards model.[11] Aklonis and Knight[12] give a good summary of the available models, as does Rohn.[13]

The crankshaft model treats the molecule as a collection of mobile segments that have some degree of free movement. This is a very simplistic approach, yet very useful for explaining behavior. As the free volume of the chain segment increases, its ability to move in various directions also increases (Figure 5.3b). This increased mobility in either side chains or small groups of adjacent backbone atoms results in a greater compliance (lower modulus) of the molecule. These movements have been studied, and Heijboer classified β and γ transitions by their type of motions.[14] The specific temperature and frequency of this softening help drive the end use of the material.

As we move from very low temperature, where the molecule is tightly compressed, we pass first through the solid state transitions. This process is shown in Figures 5.4 (6). As the material warms and expands, the free volume increases so that localized bond movements (bending and stretching) and side chain movements can occur. This is the gamma transition, $T\gamma$, which may also involve associations with water.[15] As the temperature and the free volume continue to increase, the whole side chains and localized groups of four to eight backbone atoms begin to have enough space to move and the material starts to develop some toughness.[16] This transition, called the beta transition T_β, is not as clearly defined as we are describing here (Figures 5.4 (5)). Often it is the T_g of a secondary component in a blend or of a specific block in a block copolymer. However, a correlation with toughness is seen empirically.[17]

As heating continues, we reach the T_g or glass transition, where the chains in the amorphous regions begin to coordinate large-scale motions (Figure 5.4 (4)). One classical description of this region is that the amorphous regions have begun to melt. Since the T_g only occurs in amorphous material, in a 100% crystalline material we would see not a T_g. Continued heating bring us to the $T_\alpha*$ and T_{ll} (Figure 5.4 (3)). The former occurs in crystalline or semicrystalline polymer and is a slippage of the crystallites past each other. The latter is a movement of coordinated segments in the amorphous phase that relates to reduced viscosity. These two transitions are not accepted by everyone, and their existence is still a matter of some disagreement. Finally, we reach the melt (Figure 5.4 (2)) where large-scale chain slippage occurs and the material flows. This is the melting temperature, T_m. For a cured thermoset, nothing happens after the T_g until the sample begins to burn and degrade because the cross-links prevent the chains from slipping past each other.

This quick overview gives us an idea of how an idealized polymer responds. Now let us go over these transitions in more detail with some examples of their applications. The best general collection of this information is still McCrum's 1967 text.[10]

5.3 SUB-T_g TRANSITIONS

The area of sub-T_g or higher-order transitions has been heavily studied,[18] as these transitions have been associated with mechanical properties. These transitions can

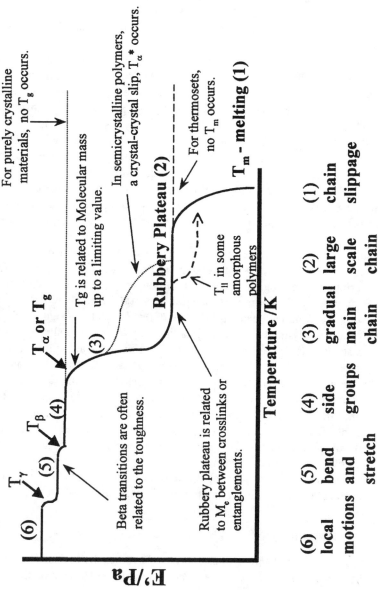

FIGURE 5.4 Idealized temperature scan of a polymer: Starting at low temperature the modulus decreases as the molecules gain more free volume, resulting in more molecular motion. This shows main curve as divided into six regions which correspond to local motions (6), bond bending and stretching (5), movements in the side chain or adjacent atoms in the main chain (4), the region of the T_g (3), coordinated movements in the amorphous portion of the chain (2), and the melting region (1). Transitions are marked as described in the text.

sometimes be seen by DSC and TMA, but they are normally too weak or too broad for determination by these methods. DMA, Dielectric Analysis (DEA), and similar techniques are usually required.[19] Some authors have also called these types of transitions second-order transitions to differentiate them from the primary transitions of T_m and T_g, which involve large sections of the main chains.[20] Boyer reviewed the T_β in 1968,[21] and pointed out that while a correlation often exists, the T_β is not always an indicator of toughness. Bershtein has reported that this transition can be considered the "activation barrier" for solid-phase reactions, deformation, flow or creep, acoustic damping, physical aging changes, and gas diffusion into polymers, as the activation energies for the transition and these processes are usually similar.[22] The strength of these transitions is related to how strongly a polymer responded to those processes. These sub-T_g transitions are associated with the materials properties in the glassy state. In paints, for example, peel strength (adhesion) can be estimated from the strength and frequency dependence of the sub-ambient beta transition.[23] Nylon 6,6 shows a decreasing toughness, measured as impact resistance, with declining area under the T_β peak in the tan δ curve. Figure 5.5 shows the relative differences in the T_β compared to the T_g for a high-impact and low-impact nylon. It has been shown, particularly in cured thermosets, that increased freedom of movement in side chains increases the strength of the transition. Cheng and colleagues report in rigid rod polyimides that the beta transition is caused by the noncoordinated movement of the diamine groups, although the link to physical properties was not investigated.[24] Johari and colleagues have reported in both mechanical[25] and dielectric studies[26] that both the β and γ transitions in bisphenol-A-based thermosets depend on the side chains and unreacted ends, and that both are affected by physical aging and postcure. Nelson has reported that these transitions can be related to vibration damping.[27] This is also true for acoustical damping.[28] In both of these cases, the strength of the beta transition is taken as a measurement of how effectively a polymer will absorb vibrations. There is some frequency dependence involved in this, which will be discussed later in Section 5.7.

Boyer[29] and Heijboer[14] showed that this information needs to be considered with care, as not all beta transitions correlate with toughness or other properties (Figure 5.6). This can be due to misidentification of the transition or to the fact that the transition does not sufficiently disperse energy. A working rule of thumb[30] is that the beta transition must be related to either localized movement in the main chain or very large side chain movement to sufficiently absorb enough energy. The relationship of large side chain movement and toughness has been extensively studied in polycarbonate by Yee,[31] as well as in many other tough glassy polymers.[32]

Less use is made of the T_γ transitions, and they are mainly studied to understand the movements occurring in polymers. Wendorff reports that this transition in polyarylates is limited to inter- and intramolecular motions within the scale of a single repeat unit.[33] Both McCrum et al.[10] and Boyd[34] similarly limited the T_γ and T_δ to very small motions either within the molecule or with bound water. The use of what is called 2D-IR, which couples an Fourier Transform Infrared Spectrometer (FTIR) and a DMA to study these motions, is a topic of current interest.[35]

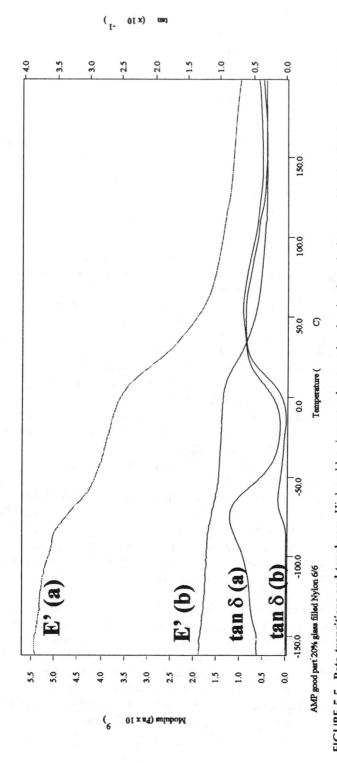

FIGURE 5.5 Beta transitions and toughness. High and low impact nylon samples showing how the beta transition is related to sample toughness as measured by impact testing. The (a) line shows a material with good impact strength by the falling dart test and the (b) line shows one with poor values by the same test.

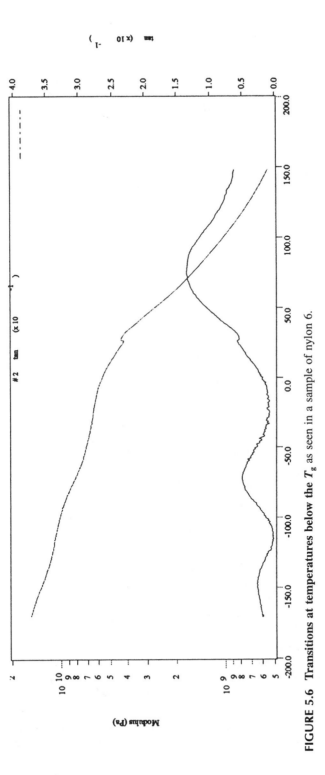

FIGURE 5.6 Transitions at temperatures below the T_g as seen in a sample of nylon 6.

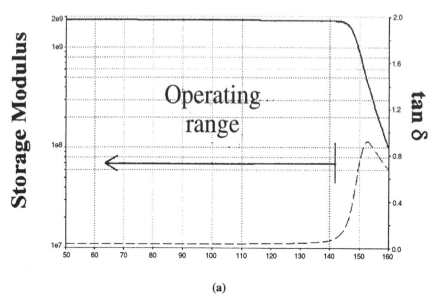

(a)

FIGURE 5.7 Operating range by DMA. Definition of operating range based on position of T_g in (a) polycarbonate, (b) epoxy, and (c) polypropylene.

5.4 THE GLASS TRANSITION (T_g OR T_α)

As the free volume continues to increase with increasing temperature, we reach the glass transition, T_g, where large segments of the chain start moving. This transition is also called the alpha transition, T_α. The T_g is very dependent on the degree of polymerization up to a value known as the critical T_g or the critical molecular weight.[36] Above this value, the T_g typically becomes less dependent on molecular weight. The T_g represents a major transition for many polymers, as physical properties changes drastically as the material goes from a hard glassy to a rubbery state. It defines one end of the temperature range over which the polymer can be used, often called the operating range of the polymer, and examples of this range are shown in Figure 5.7. For where strength and stiffness are needed, it is normally the upper limit for use. In rubbers and some semicrystalline materials such as polyethylene and polypropylene, it is the lower operating temperature. Changes in the temperature of the T_g are commonly used to monitor changes in the polymer such as plasticizing by environmental solvents and increased cross-linking from thermal or UV aging (Figure 5.8).

The T_g of cured materials or thin coatings is often difficult to measure by other methods, and more often than not the initial cost justification for a DMA is in measuring a hard-to-find T_g. While estimates of the relative sensitivity of DMA to DSC or DTA vary, it appears that DMA is 10 to 100 times more sensitive to the changes occurring at the T_g. The T_g in highly cross-linked materials can easily be seen long after the T_g has become too flat and diffuse to be seen in the DSC (Figure 5.9a). A highly cross-linked molding resin used for chip encapsulation was run by

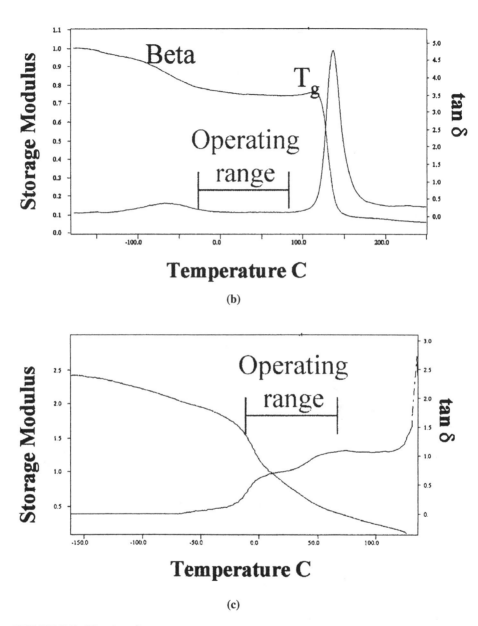

FIGURE 5.7 (*Continued*).

both methods, and the DMA is able to detect the transition after it is undetectable in the DSC. This is also a known problem with certain materials such as medical-grade urethanes and very highly crystalline polyethylenes.

The method of determining the T_g in the DMA can be a manner for disagreement, as at least five ways are in current use (Figure 5.9b). This is not unusual, as DSC has multiple methods too (Figure 5.9c). Depending on the industry standards or

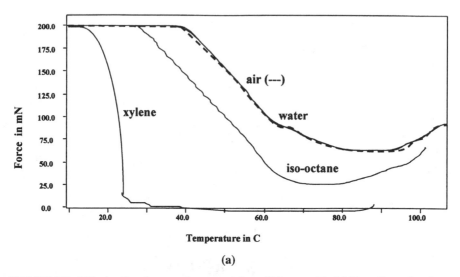

(a)

FIGURE 5.8 Effects of various environmental conditions on T_g. (a) The effect of various solvents on polypropylene fibers in air, water, iso-octane, and xylene. Note that the air and water curves overlay exactly, proving there is no effect like buoyancy or drag caused by the present of liquid supporting the probe. (b) The swelling of rubber in MEK. (c) The effect of 10 minutes immersion in saline solution on medical-grade polyurethane. (Used with permission of Rheometric Scientific, Piscataway, NJ.)

background of the operator, the peak or onset of the tan δ curve, the onset of the E' drop, or the onset or peak of the E'' curve may be used. The values obtained from these methods can differ up to 25°C from each other on the same run. In addition, a 10–20°C difference from the DSC is also seen in many materials. In practice, it is important to specify exactly how the T_g should be determined. For DMA, this means defining the heating rate, applied stresses (or strains), the frequency used, and the method of determining the T_g. For example, the sample will be run at 10°C/min under 0.05% strain at 1 Hz in nitrogen purge (20 cc/min) and the T_g determined from peak of the tan δ curve.

It is not unusual to see a peak or hump on the storage modulus directly preceding the drop that corresponds to the T_g. This is shown in Figure 5.10. This is also seen in the DSC and DTA and corresponds to a rearrangement in the molecule to relieve stresses frozen in below the T_g by the processing method. These stresses are trapped in the material until enough mobility is obtained at the T_g to allow the chains to move to a lower energy state. Often a material will be annealed by heating it above the T_g and slowly cooling it to remove this effect. For similar reasons, some experimenters will run a material twice or use a heat–cool–heat cycle to eliminate processing effects.

5.5   THE RUBBERY PLATEAU, T_α^* AND T_{ll}

The area above the T_g and below the melt is known as the rubbery plateau, and the length of it as well as its viscosity is dependent on the molecular weight between

Ntrileswell study in MEK

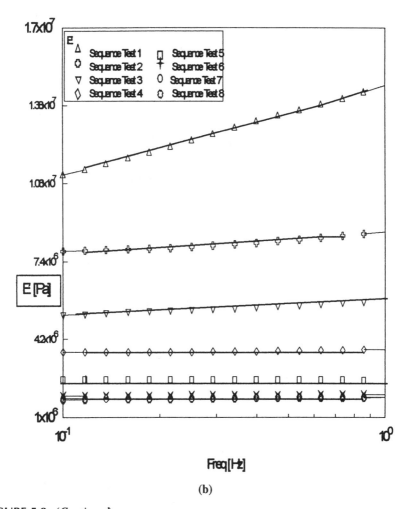

(b)

FIGURE 5.8 (*Continued*).

entanglements (M_e)[37] or cross-links. The molecular weight between entanglements is normally calculated during a stress–relaxation experiment, but similar behavior is observed in the DMA (Figure 5.11). The modulus in the plateau region is proportional to either the number of cross-links or the chain length between entanglements. This is often expressed in shear as

$$G' \cong (\rho \, RT)/M_e \qquad (5.1)$$

where G' is the modulus of the plateau region at a specific temperature, ρ is the polymer density, and M_e is the molecular weight between entanglements. In practice,

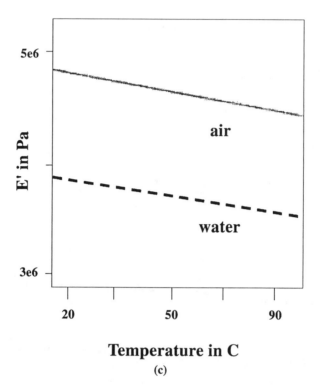

Temperature in C

(c)

FIGURE 5.8 (*Continued*).

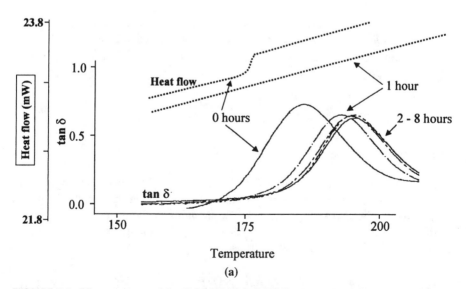

FIGURE 5.9   Measurement of the T_g by DMA and DSC. (a) The T_g of a chip encapsulation material was measured by DSC and DMA as a function of post-cure time.

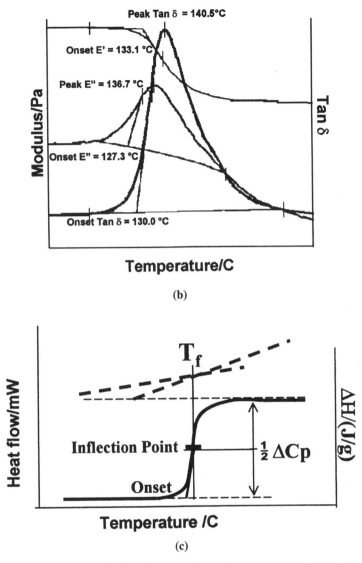

(b)

(c)

FIGURE 5.9 (*Continued*). (b) Multiple methods of determining the T_g are shown for the DMA. The temperature of the T_g varies up 10°C in this example, depending on the value chosen. Differences as great as 25°C have been reported. (c) Four of the methods used to determine the T_g in DSC are shown. The half-height and half-width methods are not included.

the relative modulus of the plateau region tells us about the relative changes in M_e or the number of cross-links compared to a standard material.

The rubbery plateau is also related to the degree of crystallinity in a material, although DSC is a better method for characterizing crystallinity than DMA.[38] This is shown in Figure 5.12. Also as in the DSC, we can see evidence of cold crystallization in the temperature range above the T_g (Figure 5.13). That is one of several

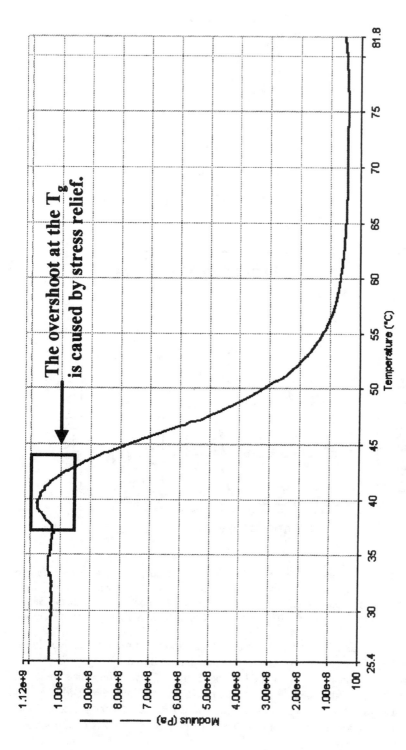

FIGURE 5.10 Stress relief at the T_g in the DMA. The overshoot is similar to that seen in the DSC and is caused by molecular rearrangements that occur due to the increased free volume at the transition.

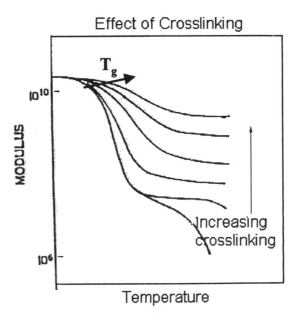

FIGURE 5.11 **Cross-linking effects.** DMA results indicating increased cross-linking by increasing T_g temperature and increasing modulus above the T_g. (Used with permission of Rheometric Scientific, Piscataway, NJ.)

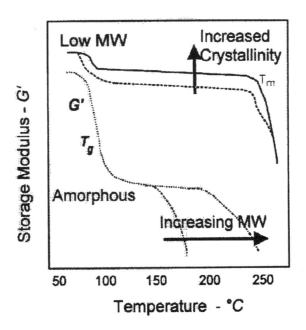

FIGURE 5.12 **Effects of crystallinity and MW on the DMA curves.** (Used with permission of Rheometric Scientific, Piscataway, NJ.)

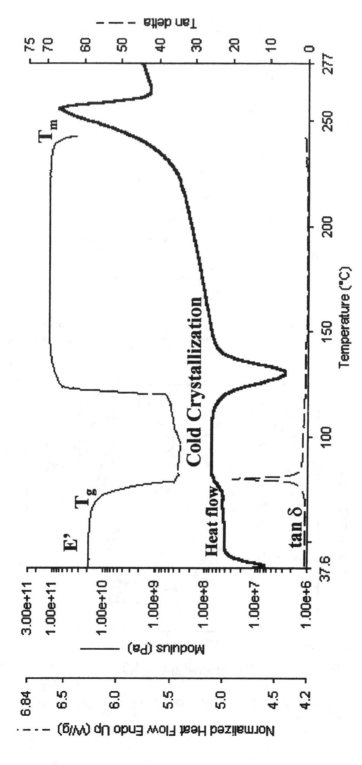

FIGURE 5.13 Cold crystallization in PET caused a large increase in the storage modulus, E', above the T_g. A DSC scan of the same material is included.

transitions that can be seen in the rubbery plateau region. This crystallization occurs when the polymer chains have been quenched (quickly cooled) into a highly disordered state. On heating above the T_g, these chains gain enough mobility to rearrange into crystallites, which causes a sometimes-dramatic increase in modulus (Figure 5.13). DSC or its temperature-modulated variant, dynamic differential scanning calorimetry (DDSC), can be used to confirm this.[39]

The alpha star transition, T_α^*, the liquid–liquid transition, T_{ll}, the heat-set temperature, and the cold crystallization peak are all transitions that can appear on the rubbery plateau. In some crystalline and semicrystalline polymers, a transition is seen here called the T_α^*.[40] Figure 5.14 shows this in a sample of polypropylene. The alpha star transition is associated with the slippage between crystallites and helps extend the operating range of a material above the T_g. This transition is very susceptible to processing-induced changes and can be enlarged or decreased by the applied heat history, processing conditions, and physical aging.[41] The T_α^* has been used by fiber manufacturers to optimize properties in their materials.

In amorphous polymers, we instead see the T_{ll}, a liquid–liquid transition associated with increased chain mobility and segment–segment associations.[42] This order is lost when the T_{ll} is exceeded and regained on cooling from the melt. Boyer reports that, like the T_g, the appearance of the T_{ll} is affected by the heat history.[43] The T_{ll} is also dependent on the number-average molecular weight, M_n, but not on the weight-average molecular weight, M_w. Bershtein suggests that this may be considered as quasi-melting on heating or the formation of stable associates of segments on cooling.[44] While this transition is reversible, it is not always easy to see, and Boyer spent many years trying to prove it was real.[45] Not everyone accepts the existence of this transition. This transition may be similar to some of the data from temperature-modulated DSC experiments showing a recrystallization at the start of the melt.[46] In both cases, some subtle changes in structure are sometimes detected at the start of melting. Following this transition, a material enters the terminal or melting region.

Depending on its strength, the heat-set temperature can also seen in the DMA. While it is normally seen in either a TMA or a CGL experiment (Figure 5.15), it will sometimes appear as either a sharp drop in modulus (E') or an abrupt change in probe position. Heat set is the temperature at which some strain or distortion is induced into polymeric fibers to change their properties, such as to prevent a nylon rug from feeling like fishing line. Since heating above this temperature will erase the texture, and you must heat polyesters above the T_g to dye them, it is of critical importance to the fabric industry. Many final properties of polymeric products depend on changes induced in processing.[47]

5.6 THE TERMINAL REGION

On continued heating, the melting point, T_m, is reached. The melting point is where the fee volume has increased so that the chains slide can past each other and the material flows. This is also called the terminal region. In the molten state, this ability to flow is dependent on the molecular weight of the polymer (Figure 5.16). The melt of a polymer material will often show changes in temperature of melting, width of

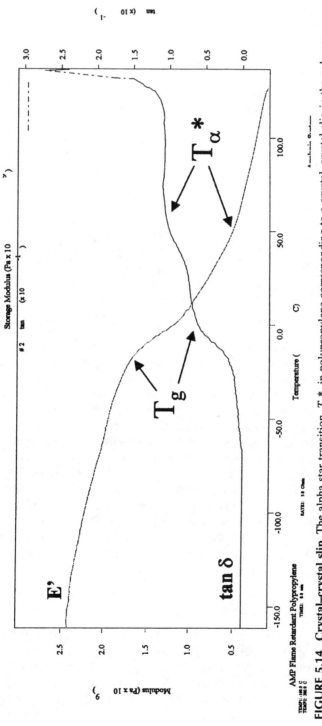

FIGURE 5.14   Crystal–crystal slip. The alpha star transition, T_α^*, in polypropylene corresponding to a crystal–crystal slip in the polymer.

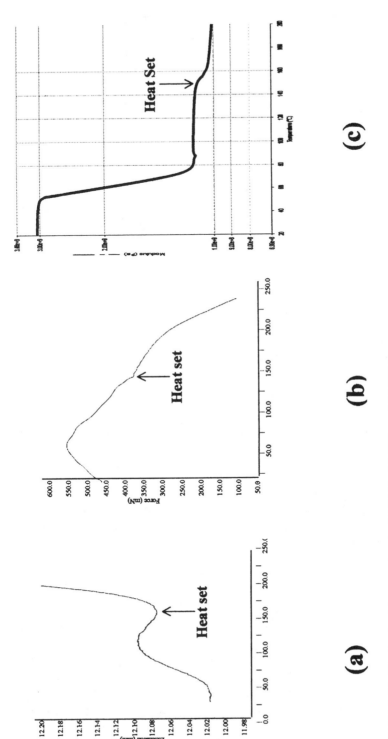

FIGURE 5.15 The heat-set temperature by (a) TMA, (b) CGL, and (c) DMA.

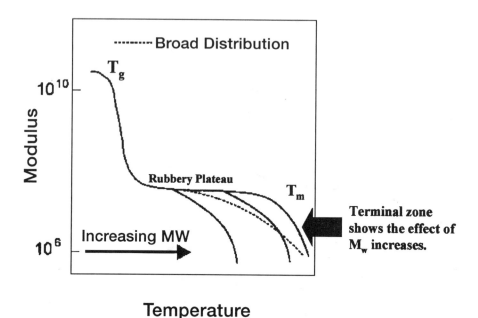

Temperature

FIGURE 5.16 The terminal zone or melting region follows the rubbery plateau and is sensitive to the M_w of the polymer. (Used with permission of Rheometric Scientific, Piscataway, NJ.)

the melting peak, and enthalpy as the material changes,[48] resulting from changes in the polymer molecular weight and crystallinity.

Degradation, polymer structure, and environmental effects all influence what changes occur. Polymers that degrade by cross-linking will look very different from those that exhibit chain scissoring. Very highly cross-linked polymers will not melt, as they are unable to flow.

The study of polymer melts and especially their elasticity was one of the areas that drove the development of commercial DMAs. Although we see a decrease in the melt viscosity as temperature increases, the DMA is most commonly used to measure the frequency dependence of the molten polymer as well as its elasticity. The latter property, especially when expressed as the normal forces, is very important in polymer processing. These topics will be discussed in detail in Chapter 7.

5.7 FREQUENCY DEPENDENCIES IN TRANSITION STUDIES

We have neglected to discuss either the choice of a testing frequency or its effect on the resulting data. While most of our discussion on frequency will be in Chapter 7, a short discussion of how frequencies are chosen and how they affect the measurement of transitions is in order. If we consider that higher frequencies induce more elastic-like behavior, we can see that there is some concern a material will act

TABLE 5.1
ASTM Tests for the DMA

D3386	CTE of Electrical Insulating Materials by TMA
D4065	Determining DMA Properties Terminology[*]
D4092	Terminology for DMA Tests
D4440	Measurement of Polymer Melts
D4473	Cure of Thermosetting Resins
D5023	DMA in Three Point Bending Tests
D5024	DMA in Compression
D5026	DMA in Tension
D5279	DMA of Plastics in Tension
D5418	DMA in Dual Cantilever
E 228-95	CTE by TMA with Silicia Dilatometer
E 473-94	Terminology for Thermal Analysis
E 831-93	CTE of Solids by TMA
E 1363-97	Temperature Calibration for TMA
E 1545-95(a)	T_g by TMA
E 1640-94	T_g by DMA
E 1824-96	T_g by TMA in Tension
E 1867-97	Temperature Calibration for DMA

[*] This standard qualifies a DMA as acceptable for all ASTM DMA Standards

This list of ASTM methods was supplied courtesy of Dr. Alan Riga. (Used with permission of Rheometric Scientific, Piscataway, NJ.)

stiffer than it really is at high test frequencies. Frequencies for testing are normally chosen by one of three methods.

The most scientific method would be to use the frequency of the stress or strain that the material is exposed to in the real world. However, this is often outside of the range of the available instrumentation. In some cases, the test method or the industry standard sets a certain frequency and this frequency is used. Ideally, a standard method like this is chosen so that the data collected on various commercial instruments can be shown to be compatible. Some of the ASTM methods for DMA are listed in Table 5.1. Many industries have their own standards, so it is important to know whether the data is expected to match a Mil-spec, an ASTM standard, or a specific industrial test. Finally, one can arbitrarily pick a frequency. This is done more often than not, so that 1 Hz and 10 rad/s are often used. As long as the data are run under the proper conditions, they can be compared to highlight material differences. This requires that frequency, stresses, and the thermal program be the same for all samples in the data set.

So what is the effect of frequency on transitions? Briefly, lowering the frequency shifts the temperature of a transition to a lower temperature (Figure 5.17a). At one time, it was suggested that multiple frequencies could be used and the T_g should then be determined by extrapolation to 0 Hz. This was never really accepted, as it represented a fairly large increase in testing time for a small improvement in accuracy. For most polymer systems, for very precise measurements, one uses a DSC. Different types of transitions also have different frequency dependencies; McCrum et al. listed many of these.[10] If one looks at the slope of the temperature dependence of transitions against frequency, one sees that in many cases the primary transitions like T_m and T_g are less dependent than the secondary transitions (Figure 5.17b). However, a perusal of McCrum's data shows this isn't always true.

5.8 PRACTICE PROBLEMS AND APPLICATIONS

In the above discussion, we have been staying mainly with either a homopolymer or with a material such as a fiber-filled composite where the filler does not show transitions. Actual commercial formulations and systems tend to be a bit messier, and I would like to address some of those issues here.

Many commercial polymers contain modifiers, and fillers[49] that are blended with the polymer to improve properties and/or to reduce costs. The concentration and presence of these additives is often best studied by other methods, but DMA lets us examine their effects on the bulk properties of the polymers. These may show up as small drops in the storage modulus curve. More likely, the effects are seen as changes to the strength and temperature of the bulk polymer. For example, changing the amount of filler or the amount of oil in peanut butter makes noticeable changes in the DMA scan as shown in Figure 5.18. Adding more filler to a rubber increases its modulus and improves the material's resistance to abrasion (Figure 5.18a). Adding oil to peanut butter softens it and makes it easier to spread (Figure 5.18b). Rigid PVC is made flexible to improve its resistance to breakage and make the tubing made from it easier to use (Figure 5.18c). Some of these effects are best seen in frequency scans and will be discussed in Chapter 7. Sometimes the polymer is added as a binder to an inorganic material, like the magnetic particles used to make the magnetic coating on a videotape. This is then coated on a polymer film. The overlapping transitions are difficult to see, and the uncoated PET film's data was subtracted from the coated film to allow detection of the transitions of the binder (Figure 5.18d).

Many polymers contain a second or third polymer as either a blend (a physical mixture of materials) or as a copolymer (a chemical mixture). This is done to toughen a hard, brittle material by adding a quantity of a rubbery material to it. The study of "rubber-toughened" or just "toughened plastics" is a large and sophisticated area of research.[50] In the DMA scan, it is often possible to see the transitions of both the main polymer and of the toughening agent.

For copolymers[51] in the DMA, one sees effects like the T_g of the copolymer moving between the extremes of the two homopolymers in relation to the molecular concentration of the components.[52] These effects are summarized in Figure 5.19, which shows how copolymers and blends change the T_g. There is a morphological

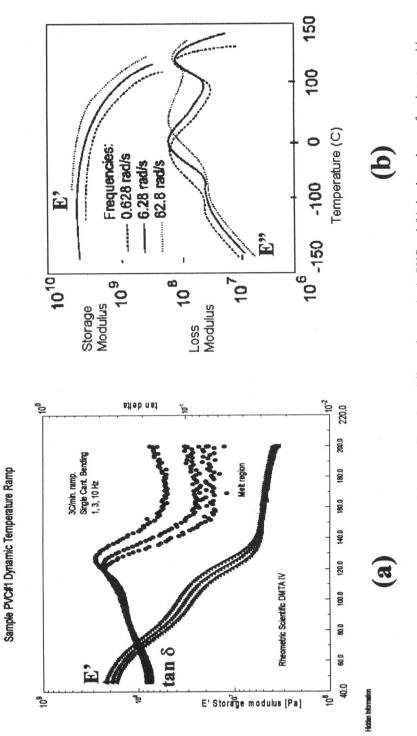

FIGURE 5.17 Frequency dependence of transitions: (a) T_g at three different frequencies in PVC and (b) the dependence of various transitions on frequency. (Used with permission of Rheometric Scientific, Piscataway, NJ.)

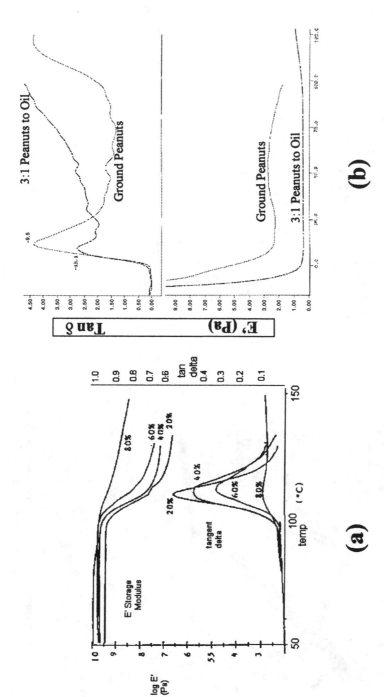

FIGURE 5.18 Effects of additives, fillers, and coatings on polymers. (a) The effect of increasing fill content on a rubber, making it harder. (Used with the permission of Rheometric Scientific, Piscataway, NJ.) (b) The effect of adding oil to peanut butter to soften it and increase its spreadability. (Data taken by Dr. Farrell Summers and used with his permission.) Rigid and flexible PVC are shown in (c), where additives decrease the stiffness of the material. Coating are used to give materials special properties like the magnetic coating applied to PET in (d). The curve for the coating is obtained by subtracting the PET curve from the coated curve.

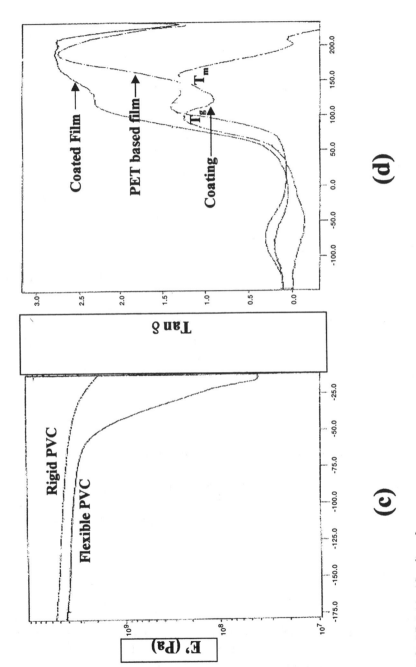

FIGURE 5.18 (*Continued*).

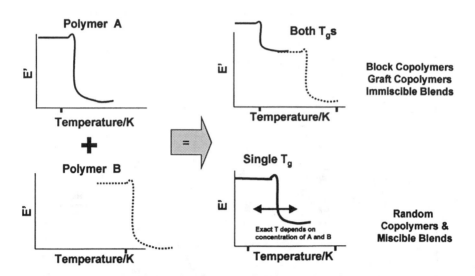

FIGURE 5.19 Blends and copolymers: how polymer blending affects the T_g.

component here too, as block copolymers will appear to be blends, if the blocks are large enough.

Certain materials such as foams represent other areas of special study. Because the bulk properties of a foam depend not only on the bulk properties of the matrix, but also on the size and spacing of the air-pockets (cells), characterization is tricky. DMA allows measurement of the matrix T_g as well as allowing characterization of the expanded foam. Macosko reports good results in studying foams in the DMA,[53] and Bessette has shown good agreement between DMA and the more traditional test methods.[54]

Finally, because DMA can give an almost instantaneous measurement of modulus, it can be used for a quick 1-minute test of a material to see its modulus and tan δ. This is often used as a quick QC tool on incoming materials or suspect products. However, it can also be used to map the modulus of a specimen to look for surface oxidation or localized embrittlement. Figure 5.20a shows a picture of a DMA system adapted to using a micro-probe (a diamond phonograph needle) to map the modulus of polyethylene parts across their wall thickness.[55] The results are shown in Figure 5.20b.

5.9 TIME-BASED STUDIES

The other part of time–temperature effects is studying how a material responds when held at constant temperature for set periods of time. This is most commonly seen in curing and post-curing studies, which will be discussed in Chapter 6. However, there are several applications where a sample will be held at a set temperature under oscillatory stress for long periods of time. Some early work using this approach was reported in Ferry,[56] where samples were run for long periods of time at varying temperatures for superposition to study age life.

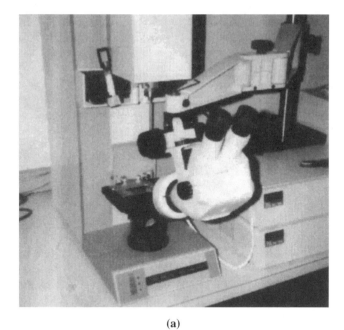

(a)

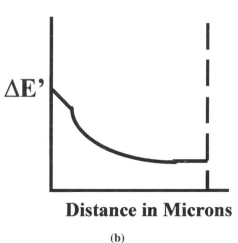

Distance in Microns

(b)

FIGURE 5.20 Micro-DMA: (a) a Perkin-Elmer DMA-7 adapted to allow use of a micro-scope with heated stage and (b) results of mapping the modulus across the thickness of a part.

However, this is not a common use of the DMA, and it is more common to hold material under constant dynamic stress at a set temperature under some sort of special conditions. This condition can simply be elevated temperature where degradation occurs or it can be a special environment, like UV light, solvents, humidity, or corrosive gases. These conditions are normally chosen to accelerate the degradation or changes seen in the final use of the material. Figure 5.21 shows the effect of elevated temperature and a cottonseed–olive oil mixture at 70°C on high-impact

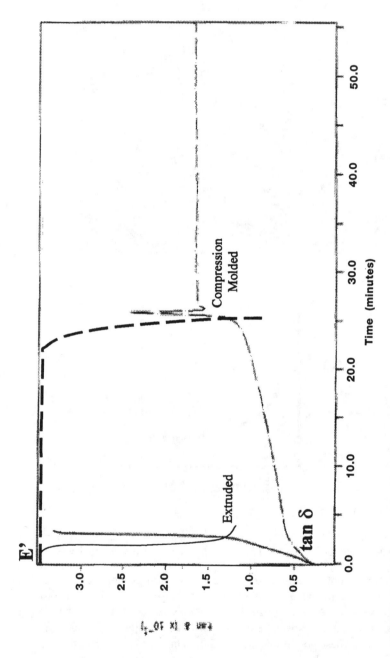

FIGURE 5.21 Degradation of HIPS in food oils by DMA: a comparison of compression-molded parts and extruded sheets. Failure is seen in both the tan δ and in the E' curves. Both samples were made from the same batch of HIPS.

polystyrene (HIPS). Other examples include medical-grade polyurethane in saline or Rykker's solution, fibers in organic solvents, oil filters in oil, Teflon piping gaskets and valves at elevated temperatures in crude oil, soft contact lenses in saline under UV, thermoset composites in high humidity, geotextile fabrics in highly acidic media, coatings in corrosive gases like H_2S, human hair coated with hair sprays at elevated temperature, and photo-curing adhesives in UV. Only a few of these approaches have been published,[57] since many of the tests are industry-specific. The ability to get failure data in these diverse conditions while exploring the changes in modulus, viscosity, and damping during the tests is a uniquely useful strength of DMA.

5.10 CONCLUSIONS

We have intentionally not covered all of the details of the effects of blends, morphology, and additives in order to limit the scope of this chapter to basic topics. Nor are we going to discuss relaxation spectra or other more advanced topics. In Chapter 8, we will discuss where you can go for more and specialized information. What is important to realize is that performing time or temperature studies in the DMA on thermoplastics or cured thermosets allows one to probe the transitions that define their properties. This is not to suggest that you should throw out your impact or Izod tester, as the promise of DMA does not always come through. It is to suggest that more transitions than the T_g are important for describing behavior in the solid state. DMA allows you to easily collect data that is difficult or costly by other means. Now, we need to consider thermosets and their study in the DMA.

NOTES

1. A. Sircar et al., in *Assignment of the Glass Transition,* R. Seyler, Ed., ASTM, Philadelphia, 1994, 293.
2. This is a fairly serious concern of a lot of users in practice, though it does not get discussed in the literature frequently. See for example R. Armstrong, *Short Course in Polymeric Fluids Rheology,* MIT, Cambridge, 1990. R. De le Garza, *Measurement of Viscoelastic Properties by Dynamic Mechanical Analysis,* Masters thesis, University of Texas at Austin, 1994. R. Hagan, *Polymer Testing,* 13, 113, 1994.
3. S. Goodkowski and B. Twombly, *Thermal Application Notes,* Perkin-Elmer, Norwalk, 56, 1994.
4. P. Flory, *Principles of Polymer Chemistry,* Cornell University Press, Ithaca, NY, 1953.
5. R. Bird, C. Curtis, R. Armstrong, and O. Hassenger, *Dynamics of Polymer Fluids,* vol. 1 & 2, 2nd ed., Wiley, New York, 1987.
6. J. D. Ferry, *Viscoelastic Properties of Polymers,* 3rd ed., Wiley, New York, 1980. J. J. Aklonis and W. J. McKnight, *Introduction to Polymer Viscoelasticity,* 2nd ed., Wiley, New York, 1983.
7. L. C. E. Struik, *Physical Aging in Amorphous Polymers and Other Materials,* Elsevier, New York, 1978. L. C. E. Struik, in *Failure of Plastics,* W. Brostow and R. D. Corneliussen, Eds., Hanser, New York, 1986, Ch. 11. S. Matsuoka, in *Failure of Plastics,* W. Brostow and R. D. Corneliussen, Eds., Hanser, New York, 1986, Ch. 3. S. Matsuoka, *Relaxation Phenomena in Polymers,* Hanser, New York, 1992.

8. J. D. Vrentas, J. L. Duda, and J. W. Huang, *Macromolecules,* 19, 1718, 1986.

9. W. Brostow and M. A. Macip, *Macromolecules,* 22(6), 2761, 1989.

10. N. McCrum, G. Williams, and B. Read, *Anelastic and Dielectric Effects in Polymeric Solids,* Dover, New York, 1967.

11. M. Doi and S. Edwards, *The Dynamics of Polymer Chains,* Oxford University Press, New York, 1986.

12. Aklonis and Knight, *Introduction to Viscoelasticity,* Wiley, New York, 1983.

13. C. L. Rohn, *Analytical Polymer Rheology,* Hanser-Gardener, New York, 1995.

14. Heijboer, *Intl. J. Polym. Mater.,* 6, 11, 1977.

15. N. McCrum, G. Williams, and B. Read, *Anelastic and Dielectric Effects in Polymeric Solids,* Dover, New York, 1967.

16. R. F. Boyer, *Polymer Eng. Sci.,* 8(3), 161, 1968.

17. Rohn, C. L., *Analytical Polymer Rheology,* Hanser-Gardener: New York, 1995, pp. 279–283.

18. J. Heijboer, *Intern. J. Polym. Mat.,* 6, 11, 1977. M. Mangion and G. Johari, *J. Polym. Sci.: Part B Polymer Physics,* 29, 437, 1991. G. Johari, G. Mikoljaczak, and J. Cavaille, *Polymer,* 28, 2023, 1987. S. Cheng et al., *Polym. Sci. Eng.,* 33, 21, 1993. G. Johari, *Lecture Notes in Physics,* 277, 90, 1987. R. Daiz-Calleja and E. Riande, *Rheol. Acta,* 34, 58, 1995. R. Boyd, *Polymer,* 26, 323, 1985. V. Bershtien, V. Egorov, L. Egorova, and V. Ryzhov, *Thermochim. Acta.,* 238, 41, 1994.

19. B. Twombly, *NATAS Proc.,* 20, 63, 1991.

20. C. L. Rohn, *Analytical Polymer Rheology,* Hanser-Gardener, New York, 1995. J. Heijboer, *Intl. J. Polym. Mat.,* 6, 11, 1977.

21. R. Boyer, *Polym. Eng. Sci.,* 8(3), 161, 1968.

22. V. Bershtein and V. Egorov, *Differential Scanning Calorimetery in the Physical Chemistry of Polymers,* Ellis Horwood, Chichester, U.K., 1993.

23. B. Coxton, private communication. Also see the sections on coatings at the SAMPE Proceedings for 1995 and 1996.

24. S. Cheng et al., *Polym. Sci. Eng.,* 33, 21, 1993.

25. G. Johari, G. Mikoljaczak, and J. Cavaille, *Polymer,* 28, 2023, 1987.

26. M. Mangion and G. Johari, *J. Polym. Sci: Part B Polymer Physics,* 29, 437, 1991.

27. F. C. Nelson, *Shock and Vibration Digest,* 26(2), 11, 1994. F. C. Nelson, *Shock and Vibration Digest,* 26(2), 24, 1994.

28. W. Brostow, private communication.

29. R. Boyer, *Polym. Eng. Sci.,* 8(3), 161, 1968.

30. G. Johari, *Lecture Notes in Physics,* 277, 90, 1987. J. Heijboer, *Intl. J. Polym. Mat.,* 6, 11, 1977. J. Heijboer et al., *Physics of Non-Crystalline Solids,* J. Prins, Ed., Interscience, New York, 1965. J. Heijboer, *J. Polymer Sci.,* C16, 3755, 1968. L. Nielsen et al., *J. Macromol. Sci. Phys.,* 9, 239, 1974.

31. A. Yee and S. Smith, *Macromolecules,* 14, 54, 1981.

32. G. Gordon, *J. Polymer Sci. A2,* 9, 1693, 1984.

33. J. Wendorff and B. Schartel, *Polymers,* 36(5), 899, 1995.

34. R. H. Boyd, *Polymer,* 26, 323, 1985.

35. I. Noda, *Appl. Spectroscopy,* 44(4), 550, 1990. V. Kien, *Proc. 6th Sympos. Radiation Chem.,* 6(2), 463, 1987.

36. L. H. Sperling, *Introduction to Physical Polymer Science,* 2nd ed., Wiley, New York, 1992.

37. C. Macosko, *Rheology,* VCH, New York, 1994.

38. F. Quinn et al., *Thermal Analysis,* Wiley, New York, 1994. B. Wunderlich, *Thermal Analysis,* Academic Press, New York, 1990.

39. J. Schawe, *Thermochim. Acta*, 261, 183, 1995. J. Schawe, *Thermochim. Acta*, 260, 1, 1995. J. Schawe, *Thermochim. Acta*, 271, 1, 1995. B. Wunderlich et al., *J. Thermal Analysis*, 42, 949, 1994.
40. R. H. Boyd, *Polymer*, 26, 323, 1985. R. H. Boyd, *Polymer*, 26, 1123, 1985.
41. S. Godber, private communication. M. Ahmed, *Polypropylene Fiber — Science and Technology*, Elsevier, New York, 1982.
42. A. Lobanov et al., *Polym. Sci. USSR*, 22, 1150, 1980.
43. R. Boyer, *J. Polym Sci. Part B: Polym. Physics*, 30, 1177, 1992. J. K. Gilham et al., *J. Appl. Polym. Sci.*, 20, 1245, 1976. J. B. Enns and R. Boyer, *Encyclopedia of Polymer Science*, 17, 23–47, 1989.
44. V. Bershtein, V. Egorov, L. Egorova, and V. Ryzhov, *Thermochim. Acta*, 238, 41, 1994.
45. C. M. Warner, *Evaluation of the DSC for Observation of the Liquid-Liquid Transition*, Masters thesis, Central Michigan State University, 1988.
46. B. Cassel et al., PETAN #69 DDSC, Perkin-Elmer, Norwalk, CT, 1995. W. Sichina, *NATAS Proc.*, 23, 137, 1994. B. Wunderlich, A. Boller, and Y. Jin, *J. Thermal Anal.*, 42, 307, 1994. B. Wunderlich, Modulated DSC, Univ. Tenn., Knoxville, 1994.
47. J. Dealy et al., *Melt Rheology and Its Role in Plastic Processing*, Van Nostrand Reinhold, Toronto, 1990. N. Chereminsinoff, *An Introduction to Polymer Rheology and Processing*, CRC Press, Boca Raton, FL, 1993.
48. E. Turi, Ed., *Thermal Characterization of Polymeric Materials*, Academic Press, Boston, 1981. E. Turi, Ed., *Thermal Analysis in Polymer Characterization*, Heydon, London, 1981.
49. Fillers can be quite interesting. See for example: S. K. De et al., *J. Appl. Polym. Sci.*, 48, 1089, 1993.
50. C. B. Bucknall, *Toughened Plastics*, Applied Science, London, 1977. R. Deanin et al., *Toughness and Brittleness of Plastics*, ACS, Washington, D.C., 1976.
51. S. Turley et al., *J. Polym. Sci.: Part C*, 1, 101, 1963. H. Keskkula, *Polym. Letters*, 7, 697, 1969. Y. Lipatov et al., *J. Appl. Polym. Sci.*, 47, 941, 1993. J. H. An, *J. Appl. Polym. Sci.*, 47, 305, 1993. B. Kim et al., *J. Appl. Polym. Sci.*, 47, 295, 1993. M. Wyzgoski, *J. Appl. Polym. Sci.*, 25, 1443, 1980.
52. N. G. McCrum, *Principles of Polymer Engineering*, Oxford University Press, New York, 1990, or L. H. Sperling, *Introduction to Physical Polymer Science*, 2nd ed., Wiley, New York, 1992.
53. C. Macosko, *NATAS Proc.*, 18, 271, 1989.
54. M. Bessette et al., *Polym. Process Eng.*, 3, 25, 1985.
55. B. Gohill and K. Menard, unpublished results.
56. J. D. Ferry, *Viscoelastic Properties of Polymers*, 3rd ed., Wiley, New York, 1980.
57. In solution testing: J. Sosa and K. Menard, *NATAS Proc.*, 25, 176, 1995. C. Daley et al., *NATAS Notes*, 26(2), 56, 1994. J. Rosenblatt et al., *J. Appl. Polym. Sci.*, 50, 953, 1993. E. McKaque et al., *J. Testing and Eval.*, 1(6), 468, 1973. Weathering: J. Pielichowski et al., *J. Thermal Anal.*, 43, 505, 1995. M. Amin et al., *J. Appl. Polym. Sci.*, 56, 279, 1995.

6 Time and Temperature Studies: Thermosets

This chapter will concentrate on the study of curing systems in the DMA. The fully cured material was treated in Chapter 5, as the concerns there are the same as for thermoplastics. However, the interest in studying curing behavior and curing materials may be even greater. The high sensitivity of the DMA and its ability to measure viscosity quickly make it one of the most valuable tools for studying curing systems. I personally have found it more useful even than DSC, although characterizing a thermoset without having both techniques available would be inefficient at best. In examining the applications of the DMA to thermosets, we will discuss fingerprinting materials, curing kinetics, methods of characterization like the Gillham–Enns diagram, post-cure studies, and decomposition studies. This chapter, like Chapter 5, will concentrate on methods that mainly involve the variation of time and temperature, although a few digressions will occur.

6.1 THERMOSETTING MATERIALS: A REVIEW

Thermosets are materials that change chemically on heating. This can occur in one step or in several, and those multiple steps do not need to be immediately sequential. In addition, many processes not normally considered chemical are studied the same way. Cakes, cookies, eggs, and meat gels (i.e., hot dog batter) are all curing systems.[1] Figure 6.1 shows the DMA scan of a commercial angel food cake batter (a) and an acrylate resin used as a dental material (b). Despite the great difference in materials, both curves show similar features and can be analyzed by the same approach. The same DMA techniques applied to traditional chemical studies can be applied to problems considered very different from those areas. The materials can even be in powdered form, as shown in Figure 6.1c, instead of solid disks or liquids. So when we discuss the cure profile below, it should be remembered that this applies to epoxies, foods, paints, coatings, and adhesives in a variety of forms. These materials, even after curing, may change on reheating, as shown by the two scans in Figure 6.1d. So care in analyzing them is required.

Thermosetting reactions can be classed into those that involve the loss of a molecule on reacting, the condensation resins, and those that join "mers" together without changes in the repeat structure, the addition condensation reactions.[2] This classification is based on the reaction mechanism of the polymers and is manifested in very different kinetics. Figure 6.2 shows the curing of a resin that releases water in the first stage of its cure and that doesn't lose a part of the molecule in the second. The first valley will often show noise, which appears as a very jagged curve, due to loss of water, while the material in the second valley undergoes chain growth. This noise is often smoothed out in practice and can be related to the kinetics on a mole

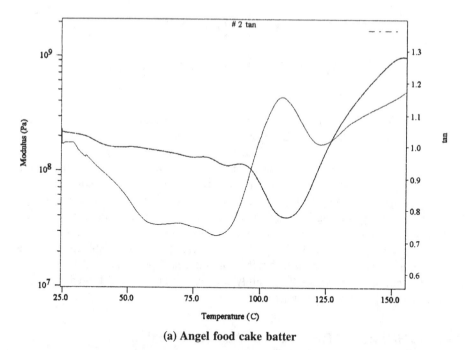

(a) Angel food cake batter

FIGURE 6.1 DMA cures. (a) DMA scan of baking commercial angel food cake batter in a DMA under the conditions specified on the box. Compare this with (b), an acrylate resin cure, to see the similar behavior of very different systems. (c) A thermosetting resin in powdered and patty (compressed) forms shows the same general behavior as other samples. The initial part of the cure is dependent on the sample preparation as shown. After the material has started melting, the runs are identical. (d) Repeat scanning of a thermoset gives different results, as the material is seldom 100% cured and the process of temperature scanning changes it.

basis. Thermogravimetric Analysis (TGA) or Thermogravimetric Analysis coupled to a Fourier transform infrared spectrometer for evolvid gas analysis (TG/IR or EGA) normally is used to collect data on the amount of water lost. I am simplifying this somewhat, and for a full development one should refer to a specialized text on polymer synthesis.[3]

Many of these curing processes are done as multi step cures, and that gives you a lot of flexibility in processing the material. A common practice is to cure the material to a given point, often to where cross-linking has just barely started, and then shape or lay-up the "α- or β-staged" resin to the final form. The alpha or beta staging refers to how long the material is cured before shipping. Beta staging is roughly twice as cured as alpha staging. This lay-up is then cured into one piece.[4] Obviously, different processes will require different degrees of staging, as staging affects the viscosity of the prepeg (an uncured resin that is impregnated onto the matrix), as shown in Figure 6.3. After the material is cured to a degree where it can support its own weight (usually 1×10^6 Pa. s), the item is removed from its mold or form and post-cured. Post-curing involves heating the free-standing piece in an oven until full mechanical strength is developed.

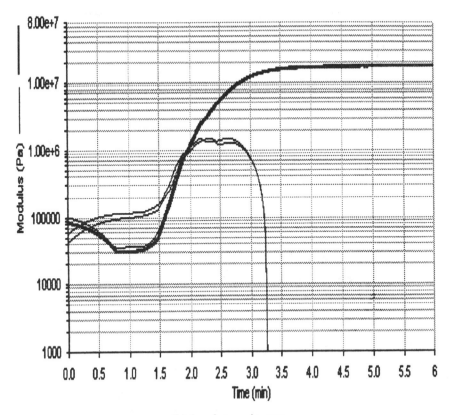

(b) Acrylate resin cure

FIGURE 6.1 (*Continued*).

Figure 6.4 shows an idealized relationship of degree of cure to T_g.[5] Initially, the material in region 1 is a monomer, and as it begins to cure it continues to act like a monomer. This continues up to about 35% of cure, when it begins to start developing polymeric properties. After a transition zone, the relationship levels off and T_g tracks well with degree of cure. The point at which the curve levels off to this slope is sometimes referred to as a critical glass transition temperature, T_g^c, and is where the materials starts showing the properties of a high polymer.[6] At some point, normally at less than 100% as measured by residual cure energy in the DSC, the rate of increase of the T_g greatly slows or stops increasing and the material has full mechanical strength. Many physical properties follow this shape of curve when plotted against molecular weight or degree of cure.[6]

For example, in some epoxy-based systems, the cure reaches a point where increased post-cure time causes little to no increase in either the modulus or T_g. At this point, increased post-curing gives no advantage and only wastes money and time.[7] In some systems, this occurs as low as 94% of complete cure when measured by the residue enthalpy of curing in the DSC. Being aware of this value and of where full mechanical strength is developed is necessary for cost-efficient process

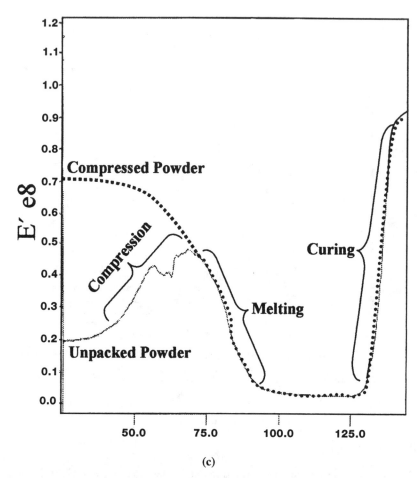

(c)

FIGURE 6.1 (*Continued*).

design. If we can develop full mechanical strength at 4 hours for the material shown in Figure 6.5, post-curing for 8 hours would only mean lost profit. As can be seen, both the T_g and the storage modulus (measured at 50°C) level out and do not increase any more after 3 to 4 hours of post-cure. This might not be true of another property such as solvent resistance or aging, and that should be checked separately.

6.2 STUDY OF CURING BEHAVIOR IN THE DMA: CURE PROFILES

The DMA's ability to give viscosity and modulus values for each point in a temperature scan allows us to estimate kinetic behavior as a function of viscosity. This has the advantage of telling us how fluid the material is at any given time, so we can determine the best time to apply pressure, what design of tooling to use, and when we can remove the material from the mold. The simplest way to

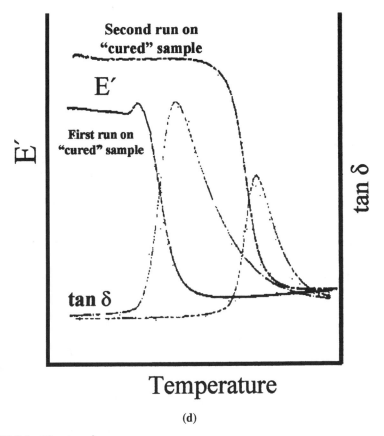

(d)

FIGURE 6.1 (*Continued*).

analyze a resin system is to run a plain temperature ramp from ambient to some elevated temperature.[8] This "cure profile" allows us to collect several vital pieces of information.

Before we analyze the cure in Figure 6.6 in more detail, we should mention that in curing studies, all three types of commercial DMAs are used. The shape of curve and the temperature of events follow the same pattern. The values for viscosity and modulus often differ greatly. Both types of forced-resonance DMAs also use samples impregnated into fabrics in techniques that are referred to as "torsion braid." There are some problems with this technique, as temperature increases will cause an apparent curing of nondrying oils as thermal expansion increases friction. However, the "soaking of resin into a shoelace," as this technique has been called, allows one to handle difficult specimens under conditions where the pure resin is impossible to run in bulk (due to viscosity or evolved volatiles). Composite materials such as graphite–epoxy composites are sometimes studied in industrial situations as the composite rather than the "neat" or pure resin because of the concern that the kinetics may be significantly different. In terms of ease of handling and sample, the composite is often easier to work with.

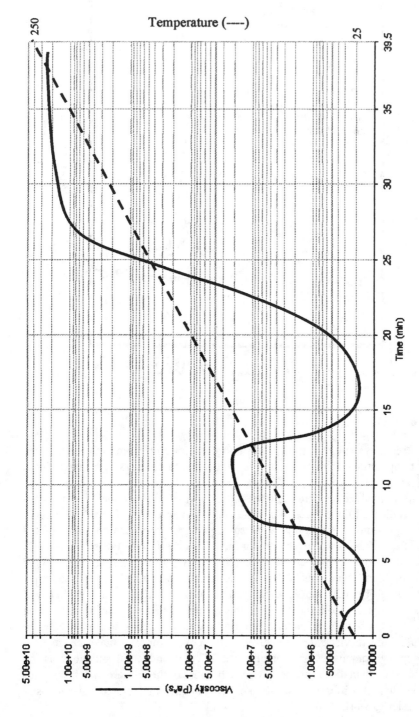

FIGURE 6.2 Cure of a two-stage condensation resin of a polyimide in the DMA. The second cure is associated with cross-linking.

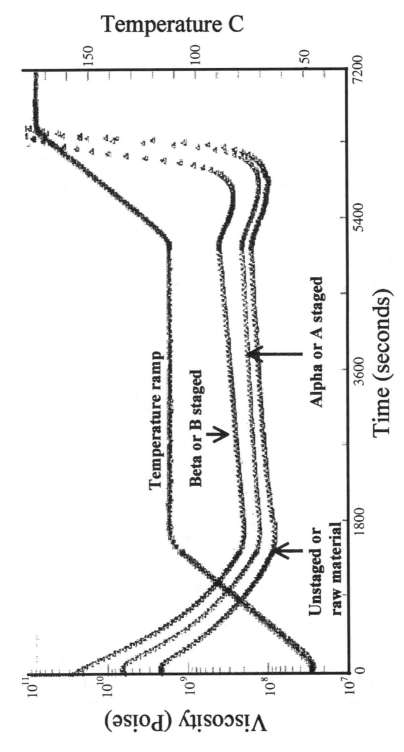

FIGURE 6.3 Effects of staging on the curing of resins. Staging is done to improve handling properties during lay-up but also changes the cure profile.

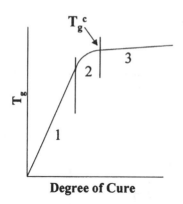

Degree of Cure

T_g^c is where the material starts exhibiting behavior characteristic of a high polymer.

- Glass Transition Temperature shows three region as a function of degree of cure or MW.
- 1- a linear relationship where the materials acts like a monomeric material
- 2-a transition zone
- 3- a region where the material has polymeric properties

FIGURE 6.4 Relationship of T_g to cure time and the stages of a cure. Note that for thermosets, it is often difficult to impossible to see the T_g by DSC in the latter half of region 3.

Another special area of concern is paints and coatings,[9] where the material is used in a thin layer. This can be addressed experimentally by either employing a braid as above or coating the material on a thin sheet of metal. The metal is often run first and its scan subtracted from the coated sheet's scan to leave only the scan of the coating. This is also done with thin films and adhesive coatings.

A sample cure profile for a commercial two-part epoxy resin is shown in Figure 6.6. From this scan, we can determine the minimum viscosity (η^*_{min}), the time to η^*_{min} and the length of time it stays there, the onset of cure, the point of gelation where the material changes from a viscous liquid to a viscoelastic solid, and the beginning of vitrification. The minimum viscosity is seen in the complex viscosity curve and is where the resin viscosity is the lowest. A given resin's minimum viscosity is determined by the resin's chemistry, the previous heat history of the resin, the rate at which the temperature is increased, and the amount of stress or stain applied. Increasing the rate of the temperature ramp is known to decrease the η^*_{min}, the time to η^*_{min}, and the gel time. The resin gets softer faster, but also cures faster. The degree of flow limits the type of mold design and when as well as how much pressure can be applied to the sample. The time spent at the minimum viscosity plateau is the result of a competitive relationship between the material's softening or melting as it heats and its rate of curing. At some point, the material begins curing faster than it softens, and that is where we see the viscosity start to increase.

As the viscosity begins to climb, we see an inversion of the E'' and E' values as the material becomes more solid-like. This crossover point also corresponds to where the tan δ equals 1 (since $E' = E''$ at the crossover). This is taken to be the gel point,[10] where the cross-links have progressed to forming an "infinitely" long net-

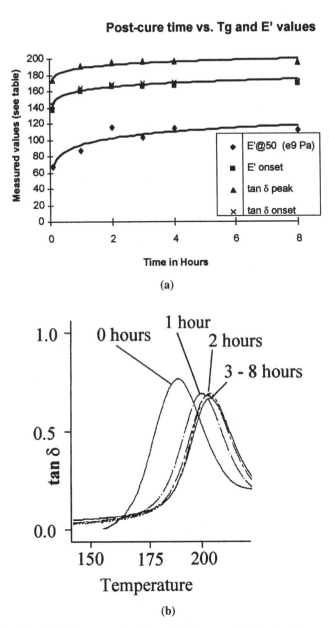

FIGURE 6.5 T_g and E' for post-cure times. (a) Data collected by DMA on chip encapsulation material plotted as time of post cure vs. measured values listed in the table. T_g was measured as the peak of the tan δ, the onset of tan δ, and the onset of the drop in E'. Storage modulus was measured at 50°C and is reported as e^9 Pa. (b) The measurement of T_g by tan δ peak values for the data in (a) is shown. All the T_g's except the 0 hour of post-cure T_g were undetectable by DSC.

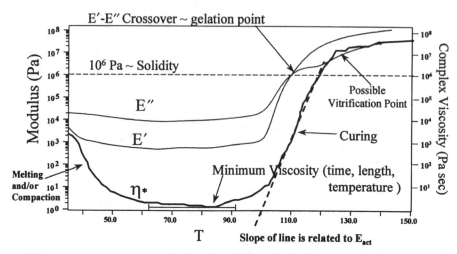

FIGURE 6.6 Analysis of a cure. The DMA cure profile of a two-part epoxy showing the typical analysis for minimum viscosity, gel time, vitrification time, and estimation of the action energy. See discussion in text.

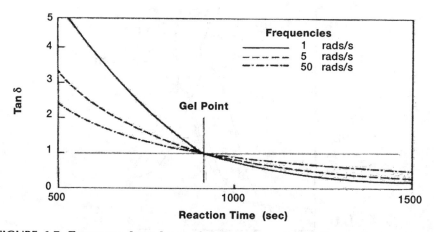

FIGURE 6.7 Frequency dependence of the gelation point. Collapse of curves run at different frequencies to a single point at the gel point. (Used with the permission of Rheometric Scientific, Piscataway, NJ.)

work across the specimen. At this point, the sample will no longer dissolve in solvent. While the gel point correlates fairly often with this crossover, it doesn't always. For example, for low initiator levels in chain addition thermosets the gel point precedes the modulus crossover.[11] A temperature dependence for the presence of the crossover has also been reported.[8] In some cases, where powder compacts and melts before curing, there may be several crossovers.[12] Then, the one following the η^*_{min} is the one of interest. Some researchers believe the true gel point is best detected by measuring the frequency dependence of the crossover point.[13] This is done by either by multiple runs at different frequencies or by multiplexing frequencies during the cure. We then look at a series of viscosity curves vs. time. (We will come back to this again in Chapter 7.) At the gel point, the frequency dependence disappears[14] (see Figure 6.7). My own experience is that this value is only a few degrees different from the one obtained in a normal scan and not worth the additional time. During this rapid climb of viscosity in the cure, the slope for η^* increase can be used to calculate an estimated E_{act} (activation energy).[15] We will discuss this below, but the fact that the slope of the curve here is a function of E_{act} is important. Above the gel temperature, some workers estimate the molecular weight, M_c, between cross-links as

$$G' = RT\rho/M_c \qquad\qquad (6.1)$$

where R is the gas constant, T is the temperature in Kelvin, and ρ is the density. At some point the curve begins to level off, and this is often taken as the vitrification point, T_{vf}.

The vitrification point is where the cure rate slows because the material has become so viscous that the bulk reaction has stopped. At this point, the rate of cure slows significantly. The apparent T_{vf}, however, is not always real: any analyzer in the world has an upper force limit. When that force limit is reached, the "topping out" of the analyzer can pass as the T_{vf}. Use of a combined technique such as DMA–DEA[16] to see the higher viscosities, or removing a sample from parallel plate and sectioning it into a flexure beam, is often necessary to see the true vitrification point (Figure 6.8). A reaction can also completely cure without vitrifying and will level off the same way. One should be aware that reaching vitrification or complete cure too quickly could be as bad as too slowly. Often a overly aggressive cure cycle will result in a weaker material, as it does not allow for as much network development, but gives a series of hard (highly cross-linked) areas among softer (lightly cross-linked) areas.

On the way to vitrification, I have marked a line at 10^6 Pa. s. This is the viscosity of bitumen[17] and is often used as a rule of thumb for where a material is stiff enough to support its own weight. This is a rather arbitrary point, but is chosen to allow the removal of materials from a mold, and the cure is then continued as a post-cure step. As an example, Table 6.1 gives the viscosities of common materials. As we shall see below, the post-cure is often a vital part of the curing process.

The cure profile is both a good predictor of performance as well as a sensitive probe of processing conditions. We will discuss the former case under Section 6.4 below and the latter as part of Section 6.7. A final note on cure profiles is that a volume change occurs during the cure.[18] This shrinkage of the resin is important and can be studied by monitoring the probe position of some DMAs as well as by TMA and dilatometry.

6.3 PHOTO-CURING

A photo-cure in the DMA is run by applying a UV light source to a sample that is held at a specific temperature or subjected to a specific thermal cycle.[19] Photo-curing is done for dental resin, contact adhesives, and contact lenses. UV exposure studies are also run on cured and thermoplastic samples by the same techniques as photo-

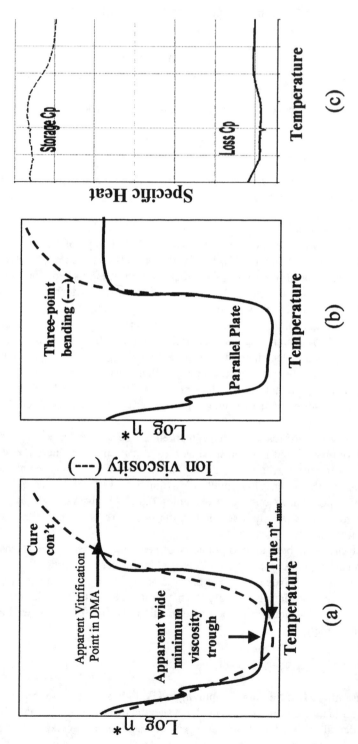

FIGURE 6.8 Masking of the vitrification point by instrument limits is shown by (a) DMA–DEA and (b) rerunning of a section of parallel plate specimen. Vitrification can also be confirmed by DDSC, as shown in (c).

TABLE 6.1
Viscosity of Common Materials

Water	0.00001 Pa s
Olive Oil	0.1 Pa s
Honey at 20°C	100 Pa s
Molasses at 20°C	1000 Pa s
Molten Polymers	10,000 Pa s
Asphalt	$1e^8$ Pa s
Solid Glass	$1e^{40}$ Pa s

curing to study UV degradation. As shown in Figure 6.9, the cure profile of a photo-cure is very similar to that of a cake or epoxy cement. The same analysis is used and the same types of kinetics developed as is done for thermal curing studies.

The major practical difficulty in running photo-cures in the DMA is the current lack of a commercially available photo-curing accessory, comparable to the photo-calorimeters on the market. One normally has to adapt a commercial DMA to run these experiments. The Perkin-Elmer DMA-7e has been successfully adapted to use quartz fixtures, a homemade heating chamber and commercial UV source triggered from the DMA's RS32 port.[20] This is a fairly easy process, and other instruments like the RheoSci DMTA Mark 4 have also been adapted.

6.4 MODELING CURE CYCLES

The above discussions are based on using a simple temperature ramp to see how a material responds to heating. In actual use, many thermosets are actually cured using more complex cure cycles to optimize the tradeoff between the processing time and the final product's properties.[21] The use of two-stage cure cycles is known to develop stronger laminates in the aerospace industry. Exceptionally thick laminates often also require multiple stage cycles in order to develop strength without porosity. As thermosets shrink on curing, careful development of a proper cure cycle to prevent or minimize internal voids is necessary.

One reason for the use of multistage cures is to drive reactions to completion. Another is to extend the minimum viscosity range to allow greater control in forming or shaping of the material. An example of a multistage cure cycle is shown in Figure 6.10. The development of a cure cycle with multiple ramps and holds would be very expensive if done with full-sized parts in production facilities. The use of the DMA gives a faster and cheaper way of optimizing the cure cycle to generate the most efficient and tolerant processing conditions.

6.5 ISOTHERMAL CURING STUDIES

Often curing is done at a constant temperature for a period of time. This is how the data needed for the kinetic models discussed in the next section are normally

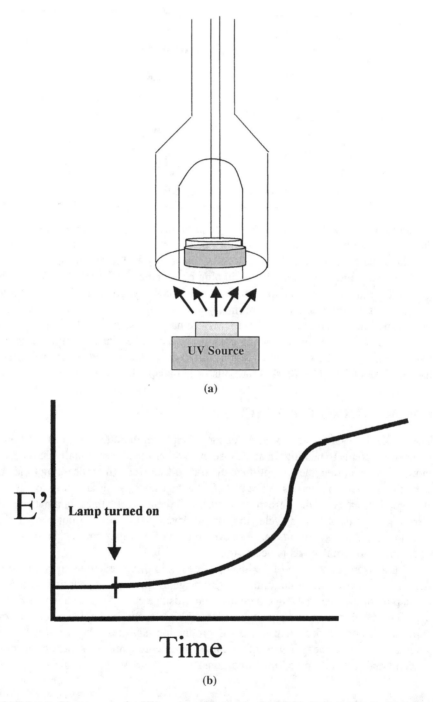

(a)

(b)

FIGURE 6.9 Photo-cure of a UV curing adhesive in the DMA. Note the similarity to the materials in Figure 6.1a and b.

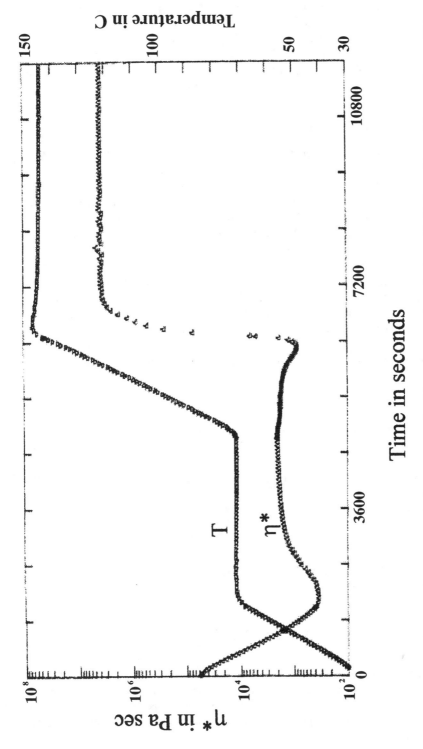

FIGURE 6.10 Multistep cure cycles: A multiple step cure cycle with two ramps and two isothermal holds is used to model processing conditions. Run on an RDA 2 by the author.

collected. It is also how rubber samples are cross-linked, how initiated reactions are run, and how bulk polymerizations are performed. Industrially, continuous processes, as opposed to batch, require an isothermal approach. Figure 6.11 shows the isothermal cure of a rubber (a) and three isothermal polymerizations (b) that were used for a kinetic study. UV light and other forms of nonthermal initiation also use isothermal studies for examining the cure at a constant temperature.

6.6   KINETICS BY DMA

Several approaches have been developed to studying the chemorheology of thermosetting systems. MacKay and Halley (Table 6.2) recently reviewed chemorheology and the more common kinetic models.[22] A fundamental method is the Williams–Landel–Ferry (WLF) model,[23] which looks at the variation of T_g with degree of cure. This has been used and modified extensively.[24] A common empirical model for curing has been proposed by Roller.[25] This method will be discussed in depth, as well as some of the variations on it.

Samples of the thermoset are run isothermally as described above, and the viscosity versus time data are plotted as shown in Figure 6.11b. This is replotted in Figure 6.12 as log η^* vs. time in seconds, where a change in slope is apparent in the curve. This break in the data indicates the sample is approaching the gel time. From these curves, we can determine the initial viscosity, η_0 and the apparent kinetic factor, k. By plotting the log viscosity vs. time for each isothermal run, we get the slope, k, and the viscosity at $t = 0$. The initial viscosity and k can be expressed as

$$\eta_0 = \eta_\infty e^{\Delta E_\eta / RT} \tag{6.2}$$

$$k = k_\infty e^{\Delta E_k / RT} \tag{6.3}$$

Combining these allows us to set up the equation for viscosity under isothermal conditions as

$$\ln \eta(t) = \ln \eta_\infty + \Delta E_\eta / RT + t k_\infty e^{\Delta E_k / RT} \tag{6.4}$$

By replacing the last term with an expression that treats temperature as a function of time, we can write

$$\ln \eta(T, t) = \ln \eta_\infty + \Delta E_\eta / RT + \int_0^t k_\infty e^{\Delta E_k / RT} dt \tag{6.5}$$

This equation can be used to describe viscosity–time profiles for any run where the temperature can be expressed as a function of time. Returning to the data plotted in Figure 6.12, we can determine the activation energies we need as follows. The plots of the natural log of the initial viscosity (determined above) vs. $1/T$ and the natural

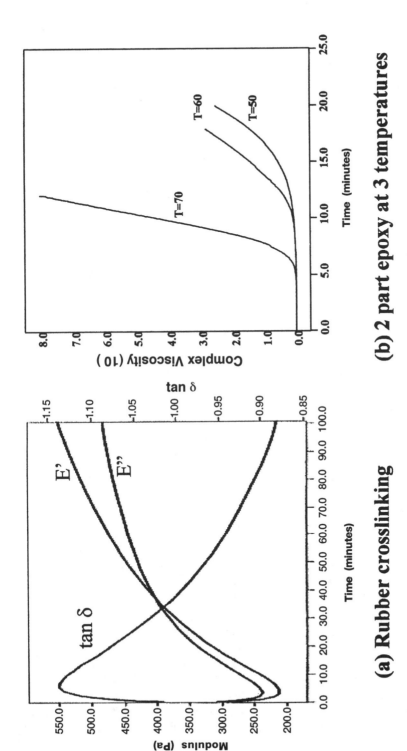

(a) Rubber crosslinking   (b) 2 part epoxy at 3 temperatures

FIGURE 6.11 Isothermal curing studies. (a) The cross-linking of a rubber sample. (Used with the permission of the Perkin-Elmer Corp., Norwalk, CT.) (b) Three isothermal runs on two-part epoxy glue.

TABLE 6.2
Chemorheological Cure Models[a]

Model	Polymer	Equation
Macosko	Epoxy, Phenolic, EDPM	$1/t_1 = C\exp(-E/RT)$
First order Iso	Epoxy	$\eta = \eta_0 \exp(\Theta t)$
First Order	Epoxy	$\eta_c = \eta(T)\exp(\phi kt)$
Empirical	Epoxy	$\ln\eta_c = \ln\eta_v + E_v/RT + k\alpha + k\alpha$
	Polyurethane	$\dfrac{\eta_c}{\eta^0} = \left(\dfrac{1+kt}{1-t/t^w}\right)^a$
Gel	Thermosets	$\dfrac{\eta_c}{\eta^0} = \left(\dfrac{\alpha^a}{\alpha^a-\alpha}\right)^{A+Ba}$
Arrhenius -1st	Epoxy	$\ln\eta_c = \ln\eta_v + E_v/RT + tk_k \exp E_k/RT$
- 1st Iso	Epoxy	$\ln\eta_c = \ln\eta_v + \dfrac{E_v}{RT} + tk_k \int \exp\left(\dfrac{E_k}{RT}\right) dt$
- nth	Epoxy	$\ln\eta_c = \ln\eta_v + \dfrac{E_v}{RT} + tk_k \int (1-\alpha)^n \exp\left(\dfrac{E_k}{RT}\right) dt$
Modified WLF		$\ln \dfrac{\eta_c(T)}{\eta_c(T_g)} = \dfrac{C_1(\alpha)[T-T_g(\alpha)]}{C_2(\alpha)+T-T_g(\alpha)}$

[a] Extracted and reprinted from Halley and MacKay, Polymer Engineering and Science, Vol. 36, No. 3, 1996, pp. 593–609, Table 3 with permission from the Society of Plastic Engineers. These models show the range of approaches used in trying to model the curing of various systems.

log of k vs. $1/T$ are used to give us the activation energies, ΔE_η and ΔE_k. Figure 6.13 shows a comparison between viscosity time profiles from actual runs and those calculated from this model. Comparison of these values to the k and ΔE of those calculated by DSC shows that this model gives larger values.[9] The DSC data is faster to obtain, but it does not include the needed viscosity information.

Several corrections have been proposed, addressing different orders of reaction[26] (the above assumes first-order) and modifications to the equations.[27]

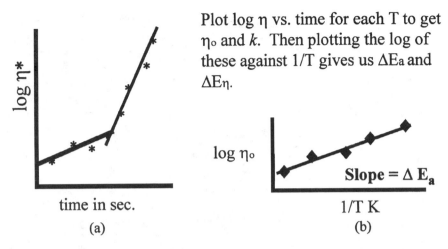

Plot log η vs. time for each T to get η₀ and k. Then plotting the log of these against 1/T gives us ΔEa and ΔEη.

FIGURE 6.12 The steps in Roller's kinetic method. Plot log η vs. time for each T to get η_o and k. Then plot the log of these against $1/T$ ΔE_a and ΔE_η. (a) shows viscosity at $t = 0$ plotted against time, and (b) shows the initial viscosity vs. $1/T$. One would also plot ln k vs. $1/T$ to obtain ΔE_η.

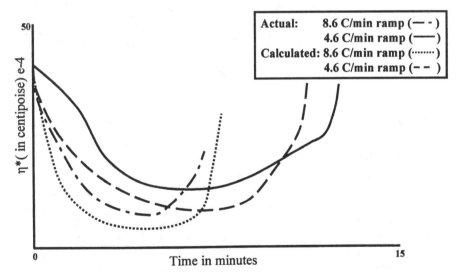

FIGURE 6.13 A comparison of Roller's predicted cure to the actual cure. (From M. Roller, *Polymer Engineering and Science,* 15, 406, 1975. With permission from the Society of Plastic Engineers.)

Many of these adjustments are reported in Roller's 1986 review of curing kinetics.[28] It is noted that these equations do not work well above the gel temperature. This same equation has been used to successfully predict the degradation of properties in thermoplastics.[29]

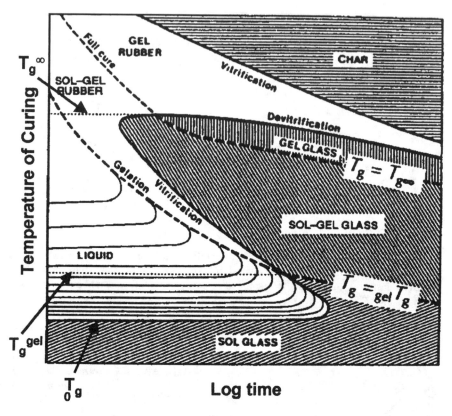

FIGURE 6.14 The Gillham–Enns or TTT diagram. (From J. K. Gillham and J. B. Enns, *Trends in Polymer Science,* 2(12), 406–419, 1994. With permission from Elsevier Science.)

6.7 MAPPING THERMOSET BEHAVIOR: THE GILLHAM–ENNS DIAGRAM

Another approach to attempt to fully understand the behavior of a thermoset was developed by Gillham[30] and is analogous to the phase diagrams used by metallurgists. The time-temperature–transition diagram (TTT) or the Gilham–Enns diagram (after its creators) is used to track the effects of temperature and time on the physical state of a thermosetting material. Figure 6.14 shows an example. These can be done by running isothermal studies of a resin at various temperatures and recording the changes as a function of time. One has to choose values for the various regions, and Gillham has done an excellent job of detailing how one picks the T_g, the glass, the gel, the rubbery, and the charring regions.[31] These diagrams are also generated from DSC data,[32] and several variants,[33] such as the continuous heating transformation and conversion-temperature-property diagrams, have been reported. Surprisingly easy to do, although a bit slow, they have not yet been accepted in industry despite their obvious utility. A recent review[34] will hopefully increase the use of this approach.

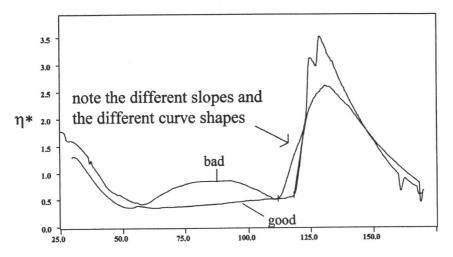

FIGURE 6.15   QC comparisons. Fingerprinting of materials for QC is often done. Good and bad hot melt adhesives were scanned using a constant heating rate cure profile.

6.8   QC APPROACHES TO THERMOSET CHARACTERIZATION

Quality control (QC) is still one of the biggest applications of the DMA in industry. For thermosets, this normally involves two approaches to examining incoming materials or checking product quality. First is the very simple approach of fingerprinting a resin. Figure 6.15 shows this for two adhesives; a simple heating run under standardized conditions allows one to compare the known good material with the questionable material. This can be done as simply as described or by measuring various quantities.

A second approach is to run the cure cycle that the material will be processed under in production and check the key properties for acceptable values. Figure 6.16 shows three materials run under the same cycle. Note the differences in the minimum viscosity, in the length and shape of the minimum viscosity plateau, the region of increasing viscosity associated with curing, and both the time required to exceed 1×10^6 Pa. s and to reach vitrification. These materials, sold for the same application, would require very different cure cycles to process. If we estimate the activation energy, E_{act}, by taking the values of η^* at various temperatures and plotting them versus $1/T$, we get very different numbers. (This is a fast way of estimating the E_{act}, where we will assume the viscosity obtained from the temperature ramp is close to the initial viscosity of the Roller method. This is not a very accurate assumption, but for materials cured under the same conditions, it works.) This indicates, as did the shape of the cures, different times are required to complete the cures. The differences in the minimum viscosity mean the material will have different flow characteristics and, for the same pressure cycle, give different thicknesses. As different times are required to reach a viscosity equal to or exceeding 10^6 Pa. s, the materials will need to be held for different times before they are solid enough to

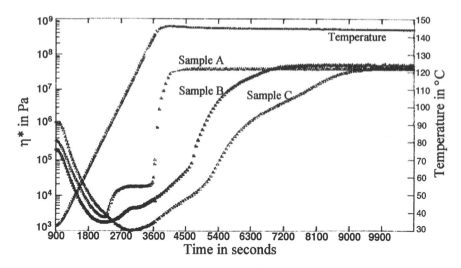

FIGURE 6.16 Comparison of materials using a standard cure cycle and analyzed as described in Figure 6.6.

hold their own weight. The overall message from this comparison is that these three materials are not interchangeable.

Another method is to convert the standard gel time test to operation in the DMA. The gel time test as done by the aerospace industry involves holding a resin sample at elevated temperature while poking it to see how long it takes to "gel." "Gel" in this case, however, means harden and this "gel" point is not the same as the gelation point estimated from the E'–E'' crossover. It is similar to the point of vitrification or complete cure on the DMA and to the vitrification point measured by DDSC. This test is the basic resin quality test for a lot of plants and is very time–consuming, as someone has to be there throughout the experiment. The same information is obtained by running an isothermal cure in the DMA. The instrument and fixtures are heated to a set temperature, often 171°C (340°F), and the run set up. The sample is then loaded quickly and a time scan done until vitrification occurs (Figure 6.17). Agreement between the DMA and manual methods is quite good (<5% in my experience) and the DMA method does not require constant monitoring.

6.9 POST-CURE STUDIES

After a material has been cured to a set level, it is often not at full strength. To allow the completion of the cure, the materials are often post-cured by heating in an oven at atmospheric pressure outside of the mold or form. This frees up expensive mold space or press time while giving a stronger laminate. This sample is normally examined for residual cure (how close to 100% cured) by DSC and checked for T_g and other transition temperatures by DMA, as described in Chapter 5. Figure 6.5 above shows the effect of post-curing on T_g. Another approach is to measure the creep of a thermoset[35] under actual use stress as a function of post cure.

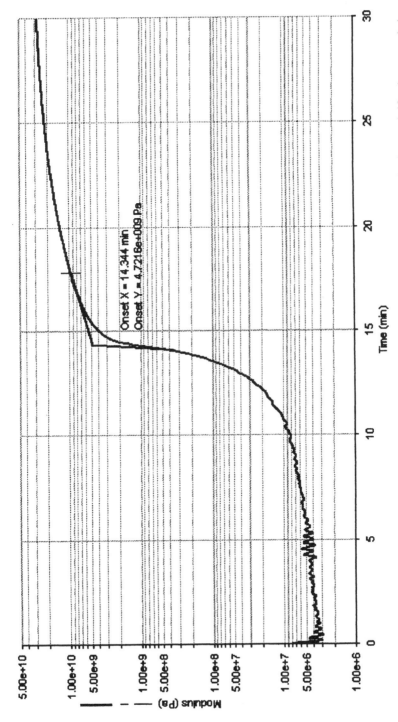

FIGURE 6.17 Results and analysis of a "gel" time test from a DMA run. Note the "gel time" here is really the time to vitrification, not gelation.

6.10   CONCLUSIONS

To summarize the above, we have seen how most curing systems can be handled by a similar approach in the DMA. The underlying similarity of the behavior of curing materials makes it fairly straightforward to investigate their behavior via the DMA. In our next chapter, we will look at the frequency-dependence of materials, which can also give us insight into the effects of the cure.

NOTES

1. C.-Y. Ma, *Thermal Analysis of Foods,* Elsevier, New York, 1990. P. Sherman et al., *Food Texture and Rheology,* Academic Press, London, 1979. C. Shoemaker et al., *Food Rheology Short Course,* S. Rheology, Boston, 1993.
2. R. B. Cassel, *Thermal Analysis Application Study # 19,* Perkin-Elmer Corp., Norwalk, CT, 1977.
3. G. Odian, *Polymer Synthesis,* Wiley, New York, 1992.
4. The area of thermoset curing and composite manufacture is very active. Some good lead references include N. G. McCrum et al., *Principles of Polymer Engineering,* Oxford Science, Oxford, 1990. L. Carlsson et al., *Experimental Characterization of Advanced Composite Materials,* Prentice-Hall, New York, 1987. J. K. Gillham et al., Eds., *Rubber-Modified Thermoset Resins,* ACS, Washington, D.C., 1984. B. Prime, in *Thermal Characterization of Polymeric Materials,* E. Turi, Ed., Academic Press, New York, 1998.
5. L. Sperling, *Introduction to Physical Polymer Science,* Wiley, New York, 1994.
6. R. Seymour and C. Carraher, *Polymer Chemistry,* Marcel Dekker, New York, 1981. R. Seymour and C. Carraher, *Structure Property Relationships in Polymers,* Plenum, New York, 1984.
7. H. Mallella et al., *ANTEC Proc.,* 40(2), 2276, 1994.
8. G. Martin et al., in *Polymer Characterization,* ACS, Washington, D.C., 1990, Ch. 12. M. Ryan et al., *ANTEC Proc.,* 31, 187, 1973. C. Gramelt, *American Laboratory,* January 26, 1984. S. Etoh et al., *SAMPE J.,* 3, 6, 1985. F. Hurwitz, *Polym. Composites,* 4(2), 89, 1983.
9. M. Roller, *Polym. Eng. Sci.,* 19, 692, 1979. M. Roller et al., *J. Coating Tech.,* 50, 57, 1978.
10. M. Heise et al., *Polym. Eng. Sci.,* 30, 83, 1990. K. O'Driscoll et al., *J. Polym. Sci.: Polym. Chem.,* 17, 1891, 1979. O. Okay, *Polymer,* 35, 2613, 1994.
11. M. Hiese, G. Martin, and J. Gotro, *Polymer Eng. Sci.,* 30 (2), 83, 1990.
12. K. Wissbrun et al., *J. Coating Tech.,* 48, 42, 1976.
13. F. Champon et al., *J. Rheology,* 31, 683, 1987. H. Winter, *Polym. Eng. Sci.,* 27, 1698, 1987. C. Michon et al., *Rheologia Acta,* 32, 94, 1993.
14. C. Michon et al., *Rheologica Acta,* 32, 94, 1993.
15. I. Kalnin, et al., *Epoxy Resins,* ACS, Washington, D.C., 1970.
16. DEA is dielectric analysis, where an oscillating electrical signal is applied to a sample. From this signal, the ion mobility can be calculated, which is then converted to a viscosity. See McCrum (op. cit.) for details. DEA will measure to significantly higher viscosities than DMA.
17. H. Barnes et al., *An Introduction to Rheology,* Elsevier, New York, 1989.
18. A. W. Snow et al., *J. Appl. Polym. Sci.,* 52, 401, 1994.

19. T. Renault et al., *NATAS Notes,* 25, 44, 1994. H. L. Xuan et al., *J. Polym. Sci.: Part A,* 31, 769, 1993. W. Shi et al., *J. Appl. Polym. Sci.,* 51, 1129, 1994.
20. J. Enns, private communication.
21. R. Geimer et al., *J. Appl. Polym. Sci.,* 47, 1481, 1993. R. Roberts, *SAMPE J.,* 5, 28, 1987.
22. P. J. Halley and M. E. MacKay, *Polym. Eng. Sci.,* 36(5), 593, 1996.
23. J. Ferry, *Viscoelastic Properties of Polymers,* 3rd ed., Wiley, New York, 1980.
24. J. Mijovic et al., *J. Comp. Met.,* 23, 163, 1989. J. Mijovic et al., *SAMPE J.,* 23, 51, 1990.
25. M. Roller, *Metal Finishing,* 78, 28, 1980. M. Roller et al., *ANTEC Proc.,* 24, 9, 1978. J. Gilham, *ACS Symp. Series,* 78, 53, 1978. M. Roller et al., *ANTEC Proc.,* 21, 212, 1975. M. Roller, *Polym. Eng. Sci.,* 15, 406, 1975. M. Roller, *Polym. Eng. Sci.,* 26, 432, 1986.
26. C. Rohn, *Problem Solving for Thermosetting Plastics,* Rheometrics, Austin, TX, 1989.
27. J. Seferis et al., *Chemorheology of Thermosetting Polymers,* ACS, Washington, D.C., 301, 1983. R. Patel et al., *J. Thermal Anal.,* 39, 229, 1993.
28. M. Roller, *Polym. Eng. Sci.,* 26, 432, 1986.
29. M. Roller, private communication, 1998.
30. J. Gillham et al., *Polym. Composites,* 1, 97, 1980. J. Enns et al., *J. Appl. Polym. Sci.,* 28, 2567, 1983. L. C. Chan et al., *J. Appl. Polym. Sci.,* 29, 3307, 1984. J. Gillham, *Polym. Eng. Sci.,* 26, 1429, 1986. S. Simon et al., *J. Appl. Polym. Sci.,* 51, 1741, 1994. G. Palmese et al., *J. Appl. Polym. Sci.,* 34, 1925, 1987. J. Enns et al., in *Polymer Characterization,* Craver, C., Ed., ACS, Washington, D.C., 1983, Ch. 2.
31. J. Gillham et al., *J. Appl. Polym. Sci.,* 53, 709, 1994. J. Enns and J. Gillham, *Trends in Polymer Sci.,* 2(12), 406, 1994.
32. A. Otero et al., *Thermochim. Acta,* 203, 379, 1992.
33. J. Gillham et al., *J. Appl. Polym. Sci.,* 42, 2453, 1991. B. Osinski, *Polymers,* 34, 752, 1993.
34. J. Enns and J. Gillham, *Trends in Polymer Sci.,* 2(12), 406, 1994.
35. R. C. Allen, *Proc. 35th Techn. Conf. On Reinforced Plastics and Composites,* 35 (26C), 1, 1980.

7 Frequency Scans

Frequency scans are the most commonly used method to study melt behavior in the DMA and, at the same time, the most neglected experiment for many users. DMA users from a rheological or polymer engineering background depend on the DMA to answer all sorts of questions about polymer melts. For many chemists and thermal analysts using DMA, the frequency scan is an ill-defined technique associated with a magical predictive method called time–temperature superposition. In this chapter, we will attempt to clear away some of the confusion and explain why the frequency dependence of a polymer is important.

7.1 METHODS OF PERFORMING A FREQUENCY SCAN

Frequency effects can be studied in various ways of changing the frequency: scanning or sweeping across a frequency range, applying a selection of frequencies to a sample, applying a complex wave form to the sample and solving its resultant strain wave, or by free resonance techniques (see Figure 7.1). Special techniques are also used to obtain collections of frequency data as a function of temperature for developing master curves and for studying the effect of frequency on temperature-driven changes in the material.

To collect frequency data, the simplest and most common approach is to hold the temperature constant and scan across the frequency range of interest. This may be done at a series of isotherms to obtain a multiplex of curves. Alternatively, one can sample a set collection of frequencies, like 1 Hz, 2.5 Hz, 5 Hz, and 10 Hz, that give a good overview on a logarithmic graph. Sampling frequencies is often performed with a simultaneous temperature scan to speed up data collection. However, since two variables are changing at the same time, there are concerns that the data are not accurate. At least two runs at different scan rates should be done so one can factor out the effects of temperature from frequency. Ideally, frequency scans should be done isothermally.

The application of a complex waveform allows very fast collection of data. By combining a set of sine waves into one wave, data can be taken for multiple frequencies in less than 30 seconds. Several approaches are used and have been reviewed by Dealy and Nelson.[1] The user should be concerned that the test is confined to the region where the Boltzmann superposition principle[2] holds for the material. Free resonance techniques,[3] discussed in Chapter 4, can also be used.

To extend the range of frequency studies to very low or high frequencies outside the instruments scanning range, data are often added from either creep or free resonance experiments. Creep data provide results at very low rates of deformation, while free resonance will provide results at the higher rates of deformation. The latter can be obtained in a stress-controlled rheometer in a recovery experiment[4] or from a specialized free resonance instrument.[3] The data from these experiments can only be added if the material acts in these tests similarly to the way it acts in a

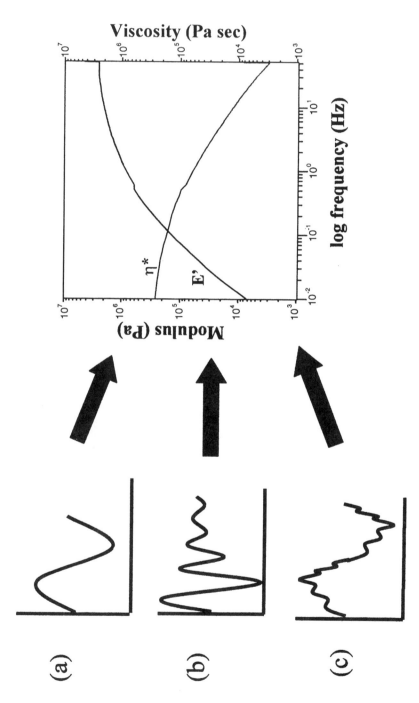

FIGURE 7.1 Methods of obtaining frequency information include (a) sequential forced frequency runs, (b) free resonance decay, and (c) complex waveforms. All can be used to generate the same types of plots.

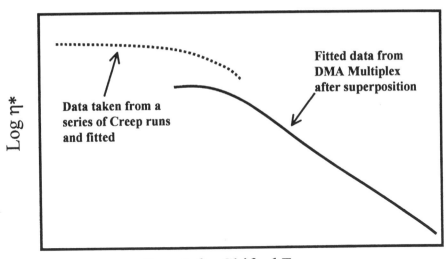

Log (af)   Shifted Frequency

FIGURE 7.2 **Frequency data from both a creep test and a DMA frequency scan** for a medical-grade polyurethane. Note that the creep data does not lie on the same curve as the frequency data obtained from a DMA run.

dynamic scan. This is not always true, as shown in Figure 7.2 for polyurethane. However, often the nature of the material is simple enough that this does work.

7.2   FREQUENCY AFFECTS ON MATERIALS

If you remember back in Chapter 2, we discussed how a fluid or polymer melt response is to strain rate (Figure 7.3) rather than to the amount of stress applied. The viscosity is one of the main reason why people run frequency scans. As the stress–strain curves and the creep–recovery runs show (Figure 7.4), viscoelastic materials exhibit some degree of flow or unrecoverable deformation. The effect is strongest in melts and liquids where frequency vs. viscosity plots are the major application of DMA.

Figure 7.5 shows a frequency scan on a viscoelastic material. In this example, the sample is a rubber above the T_g in three-point bending, but the trends and principles we discuss will apply to both solids and melts. We have plotted the storage modulus and complex viscosity on log scales against the log of frequency. Let's examine the curve and see what it tells us. First of all we should note that in analyzing the frequency scans we will be looking at trends and changes in the data, not for specific peaks or transitions.

Starting with the viscosity curve, η^*, we see at low frequency a fairly flat region called the zero shear plateau.[5] This is where the polymer exhibits Newtonian behavior and its viscosity is dependent on MW, not the strain rate. The viscosity of this plateau has been shown to be experimentally related to the molecular weight for Newtonian fluid:

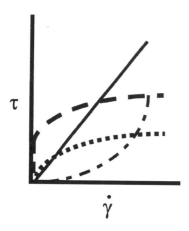

- Newtonian behavior (—) is linear and the viscosity is independent of rate.
- Pseudoplastic (•••) fluids get thinner as shear increases.
- Dilatant Fluids (– - –) increase their viscosity as shear rates increase.
- Plastic Fluids (---) have a yield point with pseudoplastic behavior.
- Thixotrophic and rheopectic fluids show viscosity-time nonlinear behavior. For example, the former shear thin and then reform its gel structure.

FIGURE 7.3 The response of a fluid to strain rate. Polymeric fluids and melts normally show deviations from Newtonian behavior.

$$\eta \propto cM_v^1 \tag{7.1}$$

for cases where the molecular weight, M_v, is less than the entanglement molecular weight, M_e, and for cases where M_v is greater than M_e,

$$\eta \propto cM_v^{3.4} \tag{7.2}$$

where η_o is the viscosity of the initial Newtonian plateau, c is a material constant, and M_v the viscosity average molecular weight. The relationship can also be written in general as replacing the exponential term with the Mark–Houwink constant, a. Similar relationships have been found for solids using different constants. This approach is used as a simple method of approximating the molecular weight of a polymer. The value obtained is closest to the viscosity average molecular obtained by osmometry.[6] In comparison with the weight average data obtained by gel permeation chromatography (GPC), the viscosity average molecular weight would be between the number average and weight average molecular weights, but closer to the latter.[7] We will discuss this again in Section 7.10.

This relationship was originally developed for continuous shear measurements, not dynamic measurements, and the question arises if we can apply it to DMA data. Since we said in earlier chapters that E' is not exactly Young's modulus and that you should not expect the dynamic modulus to equal the static modulus to equal the creep modulus, the question arises, is η^* equal to η? Cox and Merz found that an empirical relationship exists between complex viscosity and steady shear viscosity when the shear rates are the same.[8] The Cox–Merz rule is stated as

$$|\eta(\omega)| = \eta(\dot{\gamma})|_{\dot{\gamma}=\omega} \tag{7.3}$$

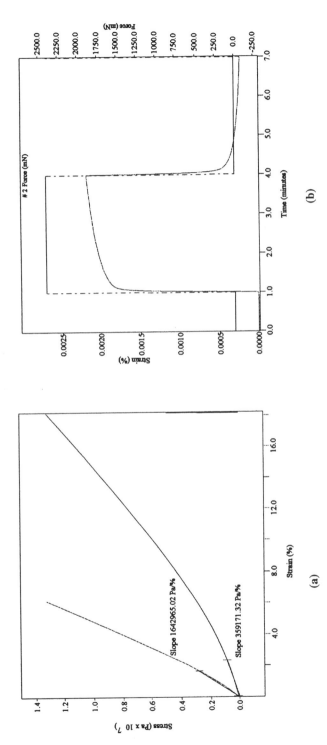

FIGURE 7.4 The effect of the viscous nature of polymers on the stress–strain and creep behavior of polymeric solids. Note both cases show nonlinearities caused by the viscoelastic, rather than elastic, nature of the materials. For (a), elastic behavior would show a flat line instead of the curved one seen, while in (b), a purely elastic material would show a square wave like the stress wave.

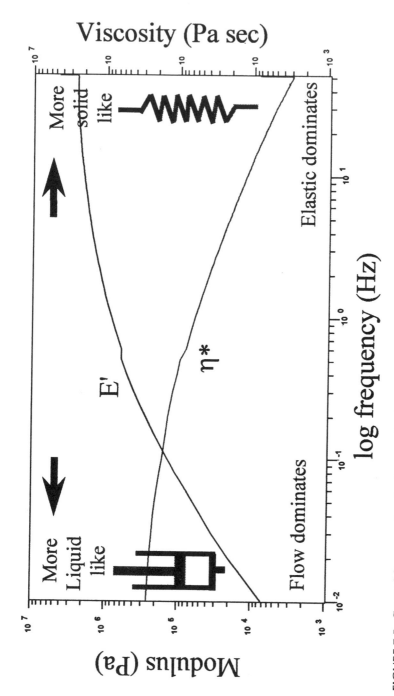

FIGURE 7.5 General frequency trends. An example of a frequency scan showing the change in a material's behavior as frequency varies. Low frequencies allow the material time to relax and respond; hence, flow dominates. High frequencies do not and elastic behavior dominates.

where η is the constant shear viscosity, η^* is the complex viscosity, ω the frequency of the dynamic test, and $d\gamma/dt$ the shear rate of the constant shear test. This rule of thumb seems to hold for most materials to within about $\pm10\%$. Another approach, which we discussed in Chapter 4, is the Gleissele mirror relationship,[9] which states the following:

$$\eta\dot{\gamma} = \eta^+(t)\Big|_{t=1/\dot{\gamma}} \qquad (7.4)$$

when $\eta^+(t)$ is the limiting value of the viscosity as the shear rate, $\dot{\gamma}$, approaches zero.

The low-frequency range is where viscous or liquid-like behavior predominates. If a material is stressed over long enough times, some flow occurs. As time is the inverse of frequency, this means we can expect materials to flow more at low frequency. As the frequency increases, the material will act in a more and more elastic fashion. Silly Putty, the children's toy, shows this clearly. At low frequency, Silly Putty flows like a liquid, while at high frequency, it bounces like a rubber ball.

This behavior is also similar to what happens with temperature changes. Remember how a polymer becomes softer and more fluid as it is heated and it goes through transitions that increase the available space for molecular motions. Over long enough time periods, or small enough frequencies, similar changes occur. So one can move a polymer across a transition by changing the frequency. This relationship is also expressed as the idea of time–temperature equivalence.[10] Often stated as "low temperature is equivalent to short times or high frequency," it is a fundamental rule of thumb in understanding polymer behavior.

As we increase the frequency in the frequency scan, we leave the Newtonian region and begin to see a relationship between the rate of strain, or the frequency, and the viscosity of the material. This region is often called the power law zone and can be modeled by

$$\eta^* \cong \eta(\dot{\gamma}) = c\dot{\gamma}^{n-1} \qquad (7.5)$$

where η^* is the complex viscosity, $\dot{\gamma}$ is the shear rate, and the exponent term n is determined by the fit of the data. This can also be written as

$$\sigma \cong \eta(\dot{\gamma}) = c\dot{\gamma}^n \qquad (7.6)$$

where σ is the stress and η is the viscosity. Other models exist, and some are given in Table 7.1. The exponential relationship is why we traditionally plot viscosity vs. frequency on a log scale. With modern curve fitting programs, the use of log–log plots has declined and is a bit anachronistic. The power law region of polymers shows the shear thickening or thinning behavior discussed in Chapter 2. This is also the region in which we find the $E'-\eta^*$ or the $E'-E''$ crossover point. As frequency increases and shear thinning occurs, the viscosity (η^*) decreases. At the same time, increasing the frequency increases the elasticity (E') increases. This is shown in Figure 7.5. The $E'-\eta^*$ crossover point is used as an indicator of the molecular weight

TABLE 7.1
Flow Models

Newtonian	$f = \eta(\dot{\gamma})$
Viscoplastic	$f - f_o = \eta(\dot{\gamma})$
Power Law	$f = k(\dot{\gamma})^n$
Power Law with Yield Stress	$f - f_o = k(\dot{\gamma})^n$
Williamson	$\eta - \eta_\infty = (\eta_o - \eta_\infty)/(1 + f/f_m)$
Cross	$\eta - \eta_\infty = (\eta_o - \eta_\infty)/(1 + \alpha(\dot{\gamma})^n)$
Carreau	$\eta - \eta_\infty = (\eta_o - \eta_\infty)/(1 + (\lambda \dot{\gamma})^a)^{(n-1)/a}$

Note: f_o is the yield stress, η_o the low shear rate viscosity, η_∞ the high shear rate viscosity, and α and n are constants.

and molecular weight distribution.[11] Changes in its position are used as a quick method of detecting changes in the molecular weight and distribution of a material. After the power law region, we reach another plateau, the infinite shear plateau.

This second Newtonian region corresponds to where the shear rate is so high that the polymer no longer shows a response to increases in the shear rate. At the very high shear rates associated with this region, the polymer chains are no longer entangled. This region is seldom seen in DMA experiments, and is usually avoided because of the damage done to the chains. It can be reached in commercial extruders and causes degradation of the polymer, which causes the poorer properties associated with regrind.

As the curve in Figure 7.5 shows, the modulus also varies as a function of the frequency. A material exhibits more elastic-like behavior as the testing frequency increases and the storage modulus tends to slope upward toward higher frequency. The storage modulus' change with frequency depends on the transitions involved. Above the T_g, the storage modulus tends to be fairly flat with a slight increase with increasing frequency as it is on the rubbery plateau. The change in the region of a transition is greater. If one can generate a modulus scan over a wide enough frequency range (Figure 7.6), the plot of storage modulus vs. frequency appears like the reverse of a temperature scan. The same time–temperature equivalence discussed above also applies to modulus, as well as compliance, tan δ, and other properties.

The frequency scan is used for several purposes that will be discussed in this chapter. One very important use, which is very straightforward, is to survey the material's response over various shear rates. This is important because many materials are used under different conditions. For example, adhesives, whether tape, Band-Aids, or hot melts, are normally applied under conditions of low frequency, and this property is referred to as tack (Figure 7.7a). When they are removed, the removal often occurs under conditions of high frequency called peel (Figure 7.7b).

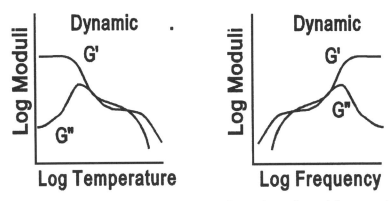

FIGURE 7.6 **Temperature and frequency scans.** Comparison of a modulus scan taken by scanning at various frequencies and by varying temperature. This relationship is called time–temperature equivalency and is discussed later in the chapter. (Used with the permission of Rheometric Scientific, Piscataway, NJ.)

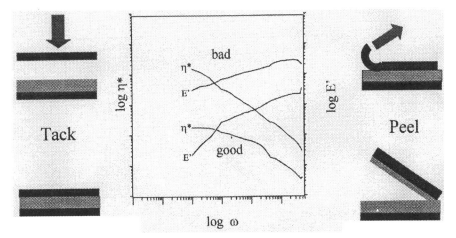

FIGURE 7.7 **Tack and peel** represent two properties that depend on opposite frequency ranges. (a) Tack is a very low frequency response involving the settling of the material into position. (b) Peel, on the other hand, is very high frequency.

Different properties are required at these regimes, and to optimize one property may require chemical changes that harm the other. Similarly, changes in polymer structure can show these kinds of differences in the frequency scan. Branching affects different frequencies differently, as shown in Figure 7.8.

For example, in a tape adhesive, we desire sufficient flow under pressure at low frequency to fill the pores of the material to obtain a good mechanical bond. When the laminate is later subjected to peel, we want the material to be very elastic so it will not pull out of the pores.[12] The frequency scan allows us to measure these properties in one scan so we can be sure that tuning one property does not degrade another. This type of testing is not limited to adhesives, as many materials see multiple frequencies in the actual use. Viscosity vs. frequency scans are used exten-

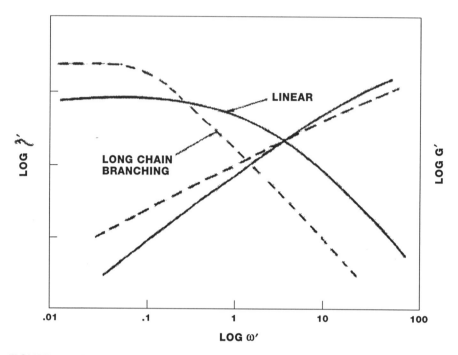

FIGURE 7.8 Comparing a long branched and a short branched polyethylene by DMA.
(Used with the permission of Rheometric Scientific, Piscataway, NJ.)

TABLE 7.2
Common Products and Their Use Frequencies

Paint Leveling	0.01 Hz
Heart Valves	1.2 Hz
Latex Gloves and Condoms	2 Hz
Plastic Hip Joints	4 Hz
Chewing, Dental Fillings	10 Hz
Contact Lenses	16 Hz
Air Bags Opening	10,000 Hz

sively to study how changes in polymer structure or formulations affect the behavior of the melt. A list of common products and the frequency of use is shown in Table 7.2. Figure 7.9 shows frequency scans on some common materials.

Another good example of why frequency scans are used is in the lay-up of graphite–epoxy composites (Figure 7.10). Composites are often laid up by hand and by mechanical tape layers in the same plant. The purchased material is designed to meet the operating requirements of both processes. However, as the material ages it becomes unsuitable for one process or the other, depending on whether it was

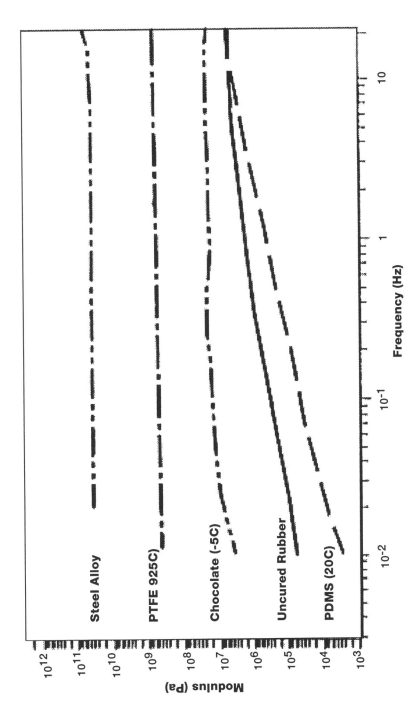

FIGURE 7.9 Frequency scans on common materials show that the modulus changes as a function of frequency more in viscoelastic materials than in elastic ones. (Used with the permission of the Perkin-Elmer Corp., Norwalk, CT.)

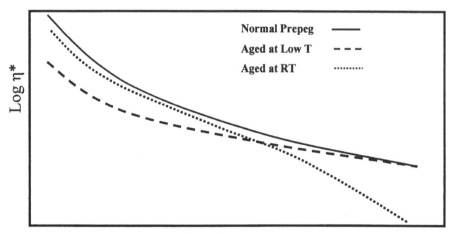

Log Frequency

FIGURE 7.10 Frequency scans on uncured three epoxy-graphite composite laminate.
Notice that different conditions show up as differences at different part of the curve. Low-frequency responses affect tack and therefore hand lay-up, while high-frequency changes affect performance in the automatic tape winders.

exposed to room temperature for too long or stored in freezers too long. The frequency scan checks both conditions in one experiment.

We need also to mention here that since we are scanning a material across a frequency range, we occasionally find conditions where the material-instrument system acts like a guitar string and begins to resonate when certain frequencies are reached. These frequencies are either the natural resonance frequency of the sample-instrument system or one of its harmonics. This is shown in Figure 7.11. Under this set of experimental conditions, the sample-instrument system is oscillating like a guitar string and the desired information about the sample is obscured. Since there is no way to change this resonance response as it is a function of the system (in fact in a free resonance analyzer we use the same effect), we will then need to redesign the experiment by changing sample dimensions or geometry to escape the problem. Using a sample with much different dimensions, which changes the mass, or changing from extension to three-point bending geometry, changes the natural oscillation frequency of the sample and hopefully solves this problem.

7.3 THE DEBORAH NUMBER

Dimensionless numbers are used to allow the comparison of material behavior from many different situations. One such number used in DMA studies is the Deborah number, defined as

$$D_e = \lambda/t = t_r/t_d \qquad (7.7)$$

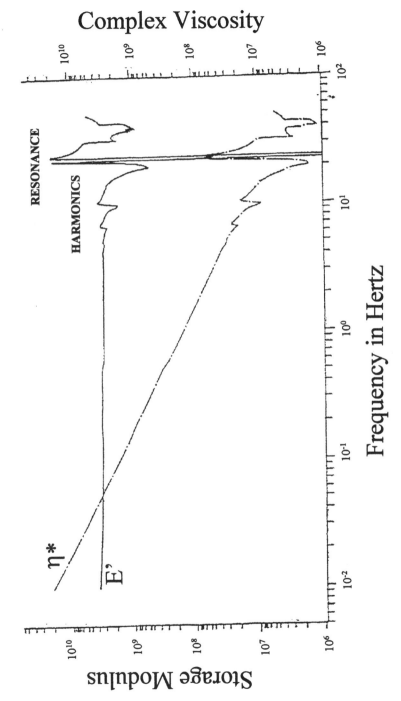

FIGURE 7.11 Free resonance occurring during a frequency scan. (Used with the permission of the Perkin-Elmer Corp., Norwalk, CT.)

where the λ is the time-scale of the material's response while **t** is the time-scale of the measurement process, which for DMA is the inverse of the frequency of measurement.[13] Rosen[13] points out that the quick estimate of λ is the relaxation time taken from a creep–recovery experiment as described in Chapter 3.4, where the time required for the material to recover to $1/e$ of the initial stress is defined as the relaxation time. Determination of an exact relaxation time for a polymer can be tricky, and it is not uncommon to plot E' versus E'' in a variation of the Cole–Cole plot to see if the polymer can really be treated as having a single relaxation time.[14,16] The t_r is the polymer's relaxation time, often taken as 1 divided by the crossover frequency in radians per second, while t_d is the deformation time. The Deborah number is used in calculations to predict polymer behavior. If

$D_e \ll 1$, the material is viscous,
$D_e \gg 1$, the material is elastic, and
$D_e \cong 1$, the material will act viscoelastically.

One use of the Deborah number is to understand how the process will affect the polymer's relaxation time. One can calculate the deformation time from the process and then see how elastic or viscous the polymer will be. By going through a process and calculating the Deborah number for each step of the process with a certain material, one can highlight areas where problems can occur.

Reiner describes how the name of the Deborah number was selected in reference to a verse from the book of Judges in the Old Testament.[15] There, in the song of Deborah, mountains are said to "flow before the Lord." The implication is that just as on our time-scale Silly Putty flows and rock is solid, on God's time-scale rock flows.

7.4 FREQUENCY EFFECTS ON SOLID POLYMERS

Both solid thermoplastics and cured thermosets are studied by various frequency methods for several reasons. First, we may be interested to see how additives or modifications affect the material over a range of frequencies. For examples, adding oils and extenders to a rubber is done to adjust properties and reduce costs. As shown in Figure 7.12, sometimes the advantage is gained in only one frequency region. By using a frequency scan, we can see if the effect occurs in frequency of actual use.

When analyzing a solid polymer, we are often looking at its transitions as a function of temperature. The frequency at which the temperature scan is run will affect the temperature of the transition. Figure 7.13a shows temperature scans run at different frequencies across a T_g. The general trend is that transitions like the glass transition move to lower temperatures as frequency decreases. In addition, the dependency of the transition on frequency is often related to the nature of the transition.[16] The sub-T_g transitions (T_β, T_γ, T_δ) are not coordinated, while the T_g requires the coordinated movement of multiple chains. Plotting the inverse of the temperature of these transitions against the log of the frequency will give different slopes for coordinated transitions than for uncoordinated ones. This is shown in Figure 7.13b. This can be exploited when investigating a polymer product for competitive analysis. By plotting the log of the frequency dependence against $1/T$

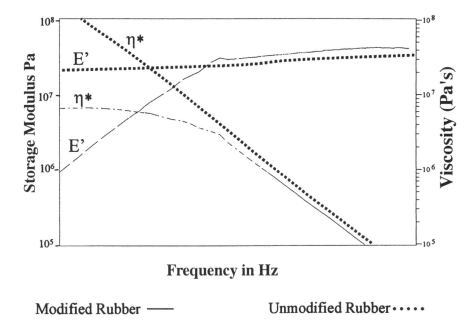

Frequency in Hz

Modified Rubber ——   Unmodified Rubber •••••

FIGURE 7.12   **The effect of adding an oil modifier to a highly cross-linked rubber.** Note that above a certain frequency, no effect of the additive is seen.

in degrees Kelvin one can estimate the activation energy of the transition. Since a T_g activation energy is about 300–400 kJ/g while a beta transition has an E_{act} of about 20–30 kJ/g, and a gamma transition has an E_{act} of 3–4 kJ/g, one can make an educated guess if the beta peak is another polymer added as a toughening agent or a side chain movement.

This dependency of transition temperature on frequency also has another implication that sometimes goes unnoticed. By changing the frequency, you can move a material through a transition, as shown in Figure 7.14. This is another reason why we need to know both the frequencies to which the material will be exposed in use and the frequency dependence of the material. If possible, even the modulus mapping discussed in Chapter 5 should be run at the use frequency.

7.5   FREQUENCY EFFECTS DURING CURING STUDIES

Cure studies are normally run at a frequency selected to allow good data to be collected or at a frequency matching that applied during the test. The application of pressure, or its increase, at various times during the minimum viscosity plateau is known to affect the final thickness of a part. Since we know that viscosity is dependent on frequency over a very wide range, the frequency at which this force is applied is also important.

By using one of the faster techniques of collecting frequency data, one can also look at the frequency dependence of the curing system. As shown in Figure 7.15, the frequency dependence reduces to zero at one point. This is taken as the point of

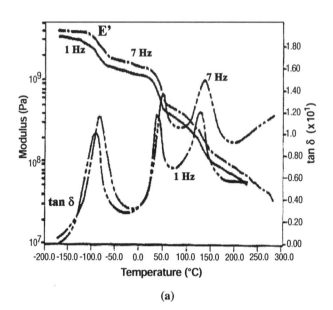

(a)

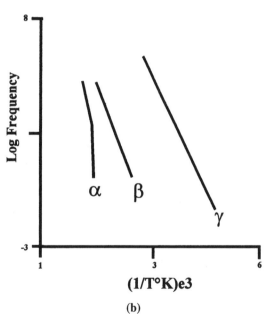

(b)

FIGURE 7.13 Effect of frequency on transitions: (a) The dependence of the T_g in poly-carbonate on frequency. (Used with the permission of Rheometric Scientific, Piscataway, NJ.) (b) Plots of the frequency dependence for polycarbonate. (Redrawn with permission from N. G. McCrum, B. E. Read, and G. Williams, *Anelastic and Dielectric Effects in Polymeric Solids,* Dover, New York, 1991.)

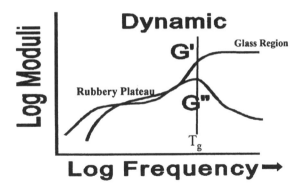

FIGURE 7.14 Frequency and transitions. The moving of a material through its transitions, such as the T_g, can be done by varying the frequency. This can be an important consideration in materials such as airbag liners, where the use frequency is close to 10,000 Hz. (Used with the permission of Rheometric Scientific, Piscataway, NJ.)

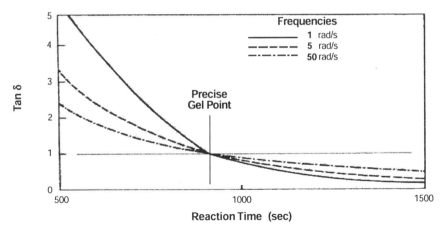

FIGURE 7.15 Gelation point. The collapse of viscosity cures run at various frequencies to one point at the gelation point. (Used with the permission of Rheometric Scientific, Piscataway, NJ.)

gelation,[17] where the network of cross-links has formed across the material. (Note the discussion in Chapter 6, where it is pointed out that the classical "gel time" test does not measure what is commonly gelation in DMA studies.) The loose network acts as a very efficient damping system at this point, and the frequency curves collapse into one. This point is normally found to be very close to the E'–E'' crossover discussed in Chapter 6, and there may be little practical advantage in this approach in many cases.

7.6 FREQUENCY STUDIES ON POLYMER MELTS

The study of polymer melts by DMA is a large enough topic to be a text in itself. In this section, we will discuss the basics and refer the reader to more advanced texts on the topic. Since almost all of the processing techniques for polymers involve the melting of these materials, this is one of the most important topics in general rheology. Dealy has written an book on melt rheology,[18] and other good texts are available. Both graduate and short courses[19] are offered that deal exclusively with this topic.

The concerns of melt rheology for the DMA operator are normally the frequency-dependence of viscosity, the elasticity or normal forces associated with the shearing of the melt, and the determination of molecular weight and distribution. The frequency dependence of a molten polymer's modulus and elasticity is determined by running a series of frequency scans as described in Section 7.1. This is done at a series of temperatures, since the viscosity will have both frequency and temperature dependencies. These data is often combined into a master-curve, as discussed below. At some point, increasing temperature or frequency will begin to irreversibly degrade the polymer by actually breaking chains. Since extrusion, injection molding, film blowing, etc., are influenced by the viscosity and modulus of the polymer for the amount of force needed to process it as well as the strength of the molten film, etc., these data are vital to a processor.

7.7 NORMAL FORCES AND ELASTICITY

One of the interesting effects in polymer extrusion is the die-swell.[20] When a polymer is processed, it springs out of the extruder and visually swells. Die-swell is the term used to describe how much a polymer melt expands when leaving the die and is critically important in die design. The swelling can be between 200–400% of the die diameter for polymers. Because of this swell, the die for extruding a square tube is slightly concave on the sides. The same effect can be easily seen in capillary rheometer studies. This requires designing dies with dimensions that are different from those of the desired product. Early work on rheology[20] reports the concern with these values. Similarly, if we stir a polymer melt or solution at high speed we see not the expected rise of the solution at the walls (caused by the centrifugal force throwing the material outwards), but that the solution instead climbs the stirrer. All of these effects are caused by the elasticity of the melt or solution.

So when we shear a material as in Figure 7.16, the entanglements of the chains cause the material to push and pull in directions normal (perpendicular) to the applied stress. This is called the normal force or the normal stresses. Normal force can be determined in certain shear rheometers by measuring how hard the polymer pushes against the top and bottom plates while sheared. One usually discusses this in terms of the normal stresses coefficients. One calculates the normal stress for each direction and then looks at the first and second normal stress coefficients. For a cubic sample where the normal stress can be called σ_x, σ_y, and σ_z, we can define the normal stress coefficients as

$$\Psi_1 = \left(\sigma_x - \sigma_y\right)\Big/\left(d\gamma/dt\right)^2 \tag{7.8}$$

$$\Psi_2 = \left(\sigma_y - \sigma_z\right)\Big/\left(d\gamma/dt\right)^2 \tag{7.9}$$

Polypropylene, Steady Rate Sweep Test, at 200°C

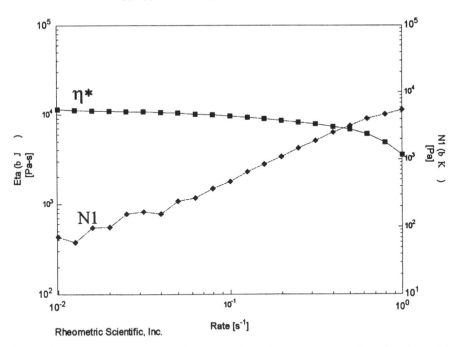

Rheometric Scientific, Inc.

FIGURE 7.16 The existence of the normal force in a polymer melt under shear. (a) When sheared the entanglements of the polymer chains cause a force to be generated normal to the direction of shear. Normal force can be measured in (b) constant shear and in (c) dynamic shear experiments. (Used with the permission of Rheometric Scientific, Piscataway, NJ.)

The second value, Ψ_2, is normally small and negative, so most of the concern is on the first normal stress coefficients, Ψ_1.

One can measure the normal stress coefficients by combining the results of two continuous shear experiments.[21] Experimentally this is done by measuring the total thrust, $\mathcal{F}$, on the lower plate in a cone and plate fixture and solving:

$$\Psi_1 = 2\mathcal{F}/\left(\pi R^2\right)\left(d\gamma/dt\right)^2 \tag{7.10}$$

where R is the radius of the plate. Then one can perform another run using parallel plates and get the difference between the normal stress coefficients as

$$\Psi_1 - \Psi_2 = \left[1/(d\gamma/dt)_R\right]\left(\mathcal{F}/\pi R^2\right)\left[2 + d\ln\left(\mathcal{F}/\pi R^2\right)/d\ln\left(d\gamma/dt\right)_R\right] \tag{7.11}$$

where R is the radius, $(d\gamma/dt)_R$ is the shear rate at the rim of the plate, and $\mathcal{F}$ the normal force (thrust against the plate). So in two experiments, we get both coefficients.[22]

Armstrong has reported that the shape of the normal force curve tracks the storage modulus shape closely, and the information collected from the E' or G' curve is often adequate to study a material's elasticity.[23] The ability of DMA to give a measurement of elasticity has, to some degree, lowered the interest in direct measurement of Ψ_1 and Ψ_2. The storage shear modulus, G', can be used to estimate Ψ_1. Under conditions where the Cox–Merz rule applies, we can assume

$$G'(\omega)/\omega^2\big|_{\omega\to 0} = N_1(\gamma)/2\gamma^2\big|_{\gamma\to 0} = \Psi_1(\gamma)/2\big|_{\gamma\to 0} \qquad (7.12)$$

where G' is the shear storage modulus, N_1 the first normal force, and Ψ_1 the first normal force coefficient. More simply put, the first normal stress difference can be expressed as

$$\Psi_1 = \left(\sigma_x - \sigma_y\right) = 2G' \qquad (7.13)$$

under conditions where the frequency is very low. One can also estimate the first normal stress different from η' and η''. Launn's rule is similar to the Cox–Merz rule above and states[24]

$$\Psi_1(d\gamma/dt) = \{2\eta''(\omega)/\omega\}\left[1 + (\eta''/\eta')^2\right]^{0.7}\Big|_{\omega=d\gamma/dt} \qquad (7.14)$$

A mirror relation has also been proposed by Gleissele:[25]

$$\Psi_1(d\gamma/dt) = \Psi_1^+(t)\big|_{t=k/d\gamma/dt} \qquad (7.15)$$

where k is a constant between 2.5 and 3 for many polymer fluids. Another area of interest is the dependence of the first normal stress difference on molecular weight, which is reported to be quite large.[26]

7.8   MASTER CURVES AND TIME–TEMPERATURE SUPERPOSITION

As previous discussed, the problem with performing frequency scans is that all instruments have a limited range, and often one wants data outside of the available range. There are a couple of approaches to addressing this problem. Experimentally one can add data collected from creep experiments at very low rates of strain (frequencies),[27] see Figure 7.17b. This is done by calculating the rate of strain of a material during its equilibrium or steady-state plateau and using the corresponding viscosity and modulus measured at those conditions. Very low frequencies can be reached by this method and added to the data collected by frequency scans. However, as creep is not the same as dynamic tests, sometimes the data are not even similar

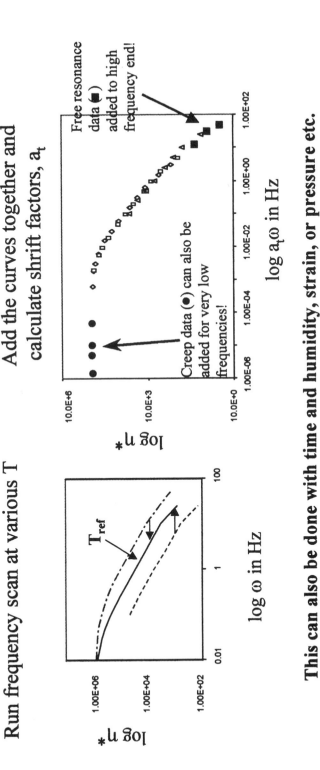

FIGURE 7.17 Time–temperature superposition. From a series of frequency scan the complex viscosity curve was collected at different temperature (a) and then added together to get a master curve (b).

as shown for the polyurethane in Figure 7.2. Due to the hard and soft segments of this polyurethane, the polymer has very different behavior under creep and DMA conditions. One can also use the results of free resonance studies to obtain higher frequencies using either a free-resonance instrument[28] or by performing a recovery experiment in a stress-controlled rheometer.[4] This was discussed in Chapter 4. There data can also be added to the frequency scan, as shown.

These approaches are relatively under-used compared to the concept of time–temperature superposition. Time–temperature superposition was described by Ferry as a "method of reduced variables."[29] Shifting a series of multiplexed frequency scans relative to a reference curve performs the superposition. This is shown in Figure 7.17a. After shifting the curves, the resultant master curve (Figure 7.17b) covers a range much greater than that of the original data.

As mentioned above, materials were studied by various techniques to obtain a series of curves often referred to as a multiplex. This is normally done to develop the master curve, which is a collection of data that have been treated so they are displayed as one curve against an axis of shifted values. This has traditionally been done using a frequency scale for the x axis and temperature as the variable to create the multiplex of curves. Using the idea of time–temperature equivalence discussed above, we can assume that the changes seen by altering the temperature are similar to those caused by frequency changes. Therefore, the data can then be superpositioned to generate one curve.

Unfortunately, current use of this technique seems to be limited to mainly time–temperature superposition. As shown by Ferry[29] and Goldman,[30] this approach can be used for a wide range of variables including humidity, degree of cure, strain, etc. Goldman's tour de force[27] gives many examples of the application of this technique to many properties of polymers. We will limit our discussion to the most commonly used approach, that of time–temperature superposition, but it is important to realize the principles can be applied to many other variables.

Various models have been developed for the shift. The most commonly used and the best know is the Williams–Landel–Ferry (WLF) model.[31] The WLF model for the shift factors is given as

$$\log a_T = \log\left(\eta/\eta_r\right) = -C_1\left(T - T_r\right)/C_2 + \left(T - T_r\right) \qquad (7.16)$$

where a_T is the shift factor, T is the temperature in degrees Kelvin, T_r is the reference temperature in degrees Kelvin, and C_1 and C_2 are material constants. The reference temperature is the temperature of the curve the data is shifted to. This is normally assumed to be valid from the T_g to 100 K above the T_g. Occasionally a vertical shift is applied to compensate for the density change of the polymer with temperature:

$$a_v = T_g\rho_g/T\rho \qquad (7.17)$$

where ρ is the density of the polymer at a temperature T.[32] After the curves are shifted, the combined curves, the master curve, can be used to predict behavior over a wide range of frequencies.

If we consider a Newtonian fluid, we can state the viscosity in terms of a flow activation energy:

$$\eta = Ae^{(E/RT)} \tag{7.18}$$

where E is the activation energy, R is the universal gas constant, and T is the temperature in degrees Kelvin. If we combine this with Equation (7.16) above, the shift factor can be written as

$$\log a_t = \log(\eta/\eta_r) = (0.434 E_{act}/R)(1/T - 1/T_r) \tag{7.19}$$

where E is the activation energy for the change in viscosity (called the flow activation energy), R is the universal gas constant, T is the temperature in degrees Kelvin of the shifted curve, T_r is the temperature in degrees Kelvin of the reference curve. Plotting the log a_t against $(1/T - 1/T_r)$ will allow us to calculate E from the slope.[33] As these values depend on the molecular parameters of the polymer, they can be used as a probe of changes in a polymer's structure. For examples, changes in molar ratio of a series of copolymer will have a corresponding change in E_{act}.[34]

Another approach to shifting curves based on free volume has been developed by Brostow and reduces to the WLF equation under certain assumptions. This equation is less limited by temperature and is

$$\ln a_t = A + B/(v_r - 1) \tag{7.20}$$

where v_r is the reduced volume of the material, calculated by dividing the molecular volume (volume per segment of polymer) of the material by a characteristic parameter called the hard core volume.[35]

Not all materials can be shifted and the term "rheologicially simple" or "rheologicial simplicity" is applied to materials that can be superpositioned. Chemical simplicity is not enough: both polyethylene and polystyrene[36] are reported as failing to superposition. Other materials like natural and synthetic rubbers are known to work quite well.

After the master curve has been generated, it can be used to predict behavior, as the basis for further manipulations to obtain relaxation or retardation spectra,[37] or to estimate the molecular weight distribution (see below). The most common uses are the prediction of behavior at a shifted frequency or as a prediction of aging. To predict the aging or long-term properties of a material, one uses the fact that frequency is measured in Hertz with units of reciprocal seconds. By inverting the curve, one can see the data against time (Figure 7.18). Note that it is the low-temperature–low-frequency data that gives the longest times after the inversion of the frequency scan. This explains the interest in measuring as low a frequency as possible. Creep data are especially valuable in this case.

If you are trying to extend the frequency range of the analyzer, the temperature chosen is normally the one at which the material will see use. It is an unfortunately

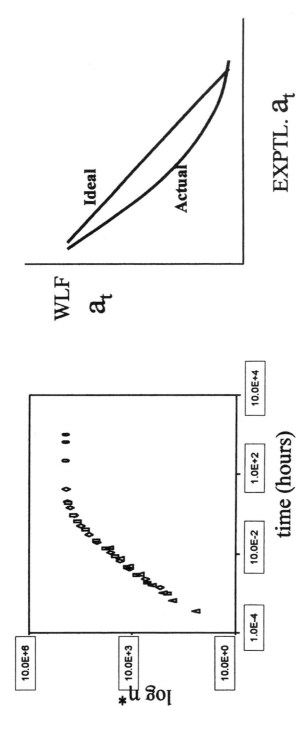

FIGURE 7.18 Conversion of frequency master curve to time. (a) The data from Figure 7.17 is converted to a time scale to predict aging. (b) The shift factors can then be plotted against a model.

common practice to ignore the theory and to shift the curves to maximize the variable you wish to study. So to obtain long times when the curve is inverted, everything is shifted to lower frequencies to obtain long times. This is done without regard to the reference temperature or the limits of the superposition model. Shifting is often done empirically, and the lines are moved up and down or straightened as necessary

Many authors have warned that superposition does not always work and is often wrong. Dealy and Wissbrum give a good discussion on this,[38] including the warning that "this assumes that all relaxation times are equally affected by temperature. This assumption is known to often be invalid." Plazek recently reviewed the approach of time–temperature superposition and pointed out that the same difficulties and failures exist today as did 20 years ago in applying this approach to various systems.[36] Some implicit assumptions exist about the mechanism of change within the sample (degradation, depolymerization, cross-linking, etc.) being the same at all temperatures. Also one assumes that different mechanisms do not occur on different sides of transitions and all rates are relatively unaffected by temperature. These are untrue for a lot of situations, and the literature on accelerated aging studies is full of case studies showing the dangers of simplistic assumptions. While this technique is a powerful tool when it works, it has been presented as a panacea, and care must be taken in its use.

7.9   TRANSFORMATIONS OF DATA

One of the advantages of frequency data is that it is possible to transform it into other forms to allow better probing of a polymer's characteristics. We have mainly used viscosity plots as examples, because the frequency data give linear plots on a log–log plot. (Log–log plots are offered on all the commercial DMAs, but this should not be taken to mean that this is the best way to handle the data.) However, the same shift factors also work for E' and E'', as all the values are calculated from the same data set. Having these data, we can generate data for master curves that would take much longer to obtain experimentally.[39] For example, we can use Ferry and Ninomiya's method to approximately calculate the equivalent stress relaxation mastercurve:[40]

$$E(t) = E'(\omega) - 0.4(E'')(0.4\omega) + 0.014(E'')(10\omega)\big|_\omega \qquad (7.21)$$

where $E(t)$ is the stress relaxation modulus, E' is the storage modulus from the dynamic experiment, ω is the dynamic test frequency, and E'' is the dynamic loss modulus. A similar equation exists for the compliance, J. This data can then be converted to a creep compliance mastercurve by

$$\int_0^1 E(t)J(t-\tau)d\tau = t \qquad (7.22)$$

and similarly for compliance. One can also convert these data to discrete viscoelastic functions such as the retardation spectra, $L(\ln \tau)$, and relaxation spectra, $H(\ln \tau)$, for the material. Figure 7.19 shows the interconversion of data. While this is beyond the scope of this book, several good references exist. These conversions, like those discussed above, are also available in software packages[41] from both instrument vendors and other sources.

7.10 MOLECULAR WEIGHT AND MOLECULAR WEIGHT DISTRIBUTIONS

It has long been known that molecular weights could be related to the polymer viscosity in the Newtonian region.[42] This was discussed above and is still used as a way to obtain the viscosity average molecular weight, M_v. The viscosity average molecular weight is larger than the number average molecular weight but slightly smaller than the weight average molecular. The viscosity average molecular weight is close enough to the latter that it responds similarly to changes in the polymer structure. The viscosity of this plateau can be related to the molecular weight for a melt by

$$\log \eta_o = cM_v^\alpha \qquad (7.23)$$

where η_o is the viscosity of the initial Newtonian plateau, c is a material constant, α is the Mark–Houwink exponent, and M_v is the viscosity average molecular weight. Above the entanglement or critical molecular weight, M_e, the value of α for melts and highly concentrated solutions is 3.4. Below that value, molecular weight is linearly related to the viscosity by a factor of 1. This is shown in Figure 7.20b. Similar relationships have been found for polymeric solids using different constants and different exponentials. If the shear rate is not in the Newtonian region, the constant a changes and at infinite shear, a becomes 1[7]. This approach is often used as a simple method of approximating the molecular weight of a polymer.

A more qualitative approach has also been used as an indicator of the relative difference in molecular weight and molecular weight distribution in polymers. Known as a rule of thumb for years, Rahalkar showed that it could be developed from the Doi–Edwards theory.[43] The crossover point between E' and E'' or between E' and η^* moves with changes in both properties (Figure 7.21a). Both points work equally well, but the theory developed for the E'–E'' crossover. As molecular weight (MW) increases, the viscosity also increases, and the crossover moves upward (toward higher viscosity). As the distribution increases, the frequency at which the material starts acting elastic increases and the point moves toward higher frequency. An example is shown in Figure 7.20b for a pair of materials. My own experience is that this approach is much more responsive to MW changes than to distribution.

The difficulty in measuring the distribution is not limited to just this approach. Measurement of the molecular weight distribution by DMA or rheology is currently a topic of discussion at many technical meetings. The Society of Plastic Engineers has had full sessions devoted to the use of rheology for quality control,[44] and these were heavily weighted toward MW and MWD measurement. Bonilla-Rios has a good overview with a detailed application in his thesis.[45] Other workers are also

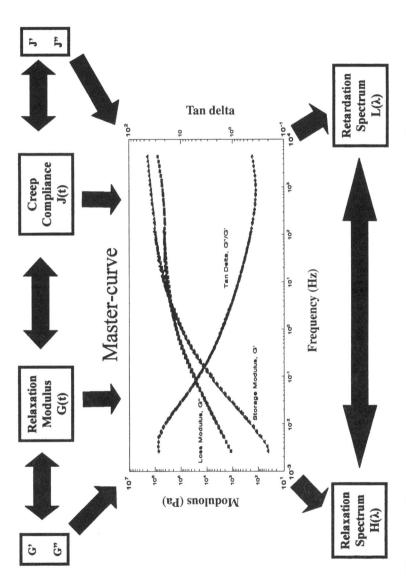

FIGURE 7.19 The interconversion of frequency scan data to different types of data. (Used with the permission of Rheometric Scientific, Piscataway, NJ.)

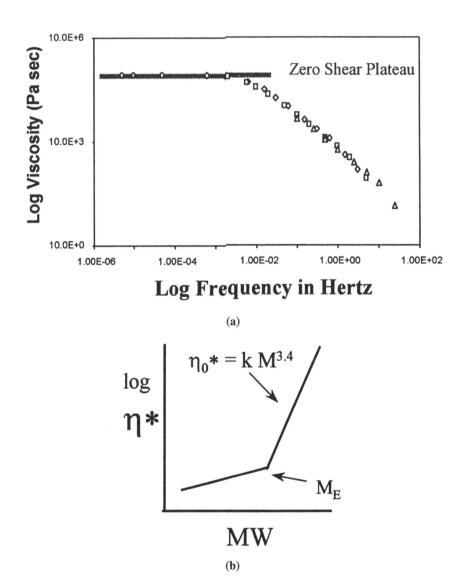

(a)

(b)

FIGURE 7.20 The relationship between the zero shear plateau's viscosity and molecular weight for a polymer melt.

excellent lead references in this area.[46] Software to perform this calculation is commercially available using an assumed normal distribution.[47] Recently, work has been published using a non-normal distribution:[48] This is of great interest, because many polymer distributions are skewed or tailed rather than normal. In fact, knowledge of how the polymer was made will often be enough to suggest whether it is high or low tailed.

The basic approach starts with the frequency-shifted master curve. Figure 7.22 shows this approach. Note the wide range of frequency data shown. We now need

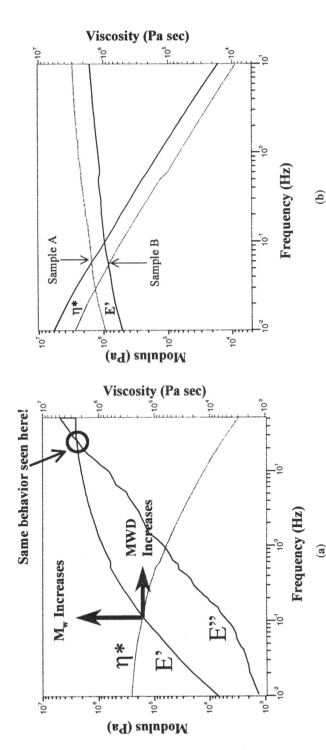

FIGURE 7.21 The crossover point (a) between either E' and E'' or between E' and η^* for a series of materials corresponds to the relative molecular weight and molecular weight distribution. This is shown for real samples in (b). (Used with the permission of Rheometric Scientific, Piscataway, NJ.)

a mixing rule. A mixing rule is a quantitative relationship that relates the observed mechanical properties of a polydisperse melt and the underlying polymer structure. This can be a relatively simple mathematical approximation to a more complex molecular theory of polydispersity and is normally supplied by the software. For example, a double reptation mixing rule like

$$G(t) = G_N \left[\int_0^\infty F^{1/2}(M,t)W(M)dM \right]^2 \quad (7.24)$$

is used in one commercial package. $G(t)$ is the relaxation modulus, which can be determined from experiments as discussed above while $F^{1/2}(M,t)$ is the monodisperse relaxation function, which represents the time-dependent fractional stress relaxation of a monodisperse polymer following a small step strain. $W(M)$ is the weight-based MWD. Physically, all components of the MWD will contribute to the modulus to some extent. The magnitude of each component's contribution to the stress will depend on the details of the interaction with the other molecules in the MWD. One normally needs to supply the plateau modulus (G_n) and $F^{1/2}(M,t)$ in addition to the above. While the plateau modulus can be obtain from the literature, some form of the relaxation function must be assumed, such as

$$F^{1/2}(M,t) = \exp\left\{ \frac{-t}{2\lambda(M)} \right\} \quad (7.25)$$

and

$$\lambda(M) = K(T)M^x \quad (7.26)$$

where $\lambda(M)$ is the characteristic relaxation time for the monodisperse system, $K(T)$ is a coefficient that depends on temperature, and the exponent x is ~3.4 for flexible polymers.

One then grinds through the mathematics, or for most of us, allows the software package to do so. For many commercially manufactured polymers, where a normal distribution is a valid assumption, good results can be obtained.

7.11 CONCLUSIONS

Frequency scans and frequency dependencies are probably the least used and the most powerful techniques in DMA. While well known among people working with melts, the average user who comes from the thermal analysis or chemistry background normally ignores them. They represent a powerful probe of material properties that should be in any testing laboratory.

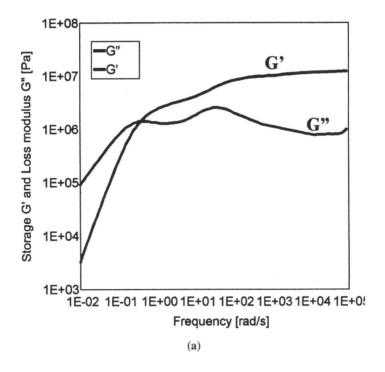

(a)

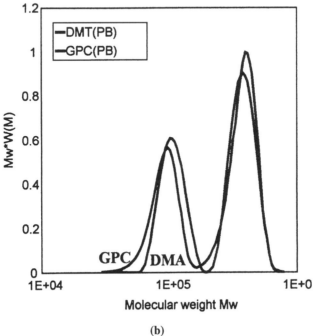

(b)

FIGURE 7.22 Determination of the molecular distribution from the frequency-shifted master curve. (Used with the permission of Rheometric Scientific, Piscataway, NJ.)

NOTES

1. J. Dealy and B. Nelson, in *Techniques in Rheological Measurement,* A. Collyer, Ed., Chapman and Hall, New York, 1993, Ch. 7.
2. I. M. Ward and D. W. Hadley, *Introduction to the Mechanical Properties of Solid Polymers,* Wiley, New York, 1993, 51–54.
3. J. Enns and J. Gilham, in *Polymer Characterization: Spectroscopic, Chromatographic, and Physical Instrumental Methods,* C. Craver, Ed., ACS, Washington, D.C., 1983, 27.
4. U. Zolzer and H-F. Eicke, *Rheologia Acta,* 32, 104, 1993.
5. C. Rohn, *Analytical Polymer Rheology,* Hanser, New York, 1995.
6. M. Miller, *The Structure of Polymers,* Reinhold, New York, 1966, 611–612.
7. S. Rosen, *Fundamantal Principles of Polymeric Materials,* Wiley Interscience, New York, 1993, 53–77, 258–259.
8. W. Cox and E. Merz, *J. Polym. Sci.,* 28, 619, 1958. P. Leblans et al., *J. Polymer Science,* 21, 1703, 1983.
9. W. Gleissele, in *Rheology,* vol. 2, G. Astarita et al., Ed., Plenum Press, New York, 1980, 457.
10. D. W. Van Krevelin, *Properties of Polymers,* Elsevier, New York, 1987, 289.
11. C. Macosko, *Rheology,* VCH Publishers, New York, 1996, 120–127.
12. This is a very simplified version of adhesion. The reader is referred to the following for a detailed discussion: L-H. Lee, Ed., *Adhesive Bonding,* Plenum Press, New York, 1991. L-H. Lee, Ed., *Fundamentals of Adhesion,* Plenum Press, New York, 1991.
13. S. Rosen, *Fundamental Principles of Polymeric Materials,* 2nd ed., Wiley, New York, 1993.
14. S. Havriliak and S.J. Hvriliak, *Dielectric and Mechanical Relaxations in Materials,* Hanser, New York, 1997.
15. M. Reiner, *Twelve Lectures on Rheology,* North Holland, Amsterdam, 1949.
16. N. McCrum, B. Read, and G. Williams, *Anelastic and Dielectric Effects in Polymeric Solids,* Dover, New York, 1995, Ch. 5, 7–14.
17. F. Champon et al., *J. Rheology,* 31, 683, 1987. H. Winter, *Polym. Eng. Sci.,* 27, 1968, 1987. C. Michon et al., *Rheologia Acta,* 32, 94, 1993.
18. J. Dealy and K. Wissbrum, *Melt Rheology and Its Role in Polymer Processing,* Van Nostrand Reinhold, Toronto, 1990.
19. Both the University of Minnesota and the Massachusetts Institute of Technology offer week-long courses.
20. R. Racin and D. Bogue, *J. Rheology,* 23, 263, 1979. B. Bagley and H. Schreiber, in *Rheology,* vol. 5, R. Eirich, Ed., Academic Press, New York, 1969, 93.
21. R. Bird, R. Armstrong, and O. Hassager, *Dynamics of Polymeric Liquids,* V1, Wiley, New York, 1987, 521–529.
22. C. Macosko, *Rheology,* VCH Publishers, New York, 1994, 205–229.
23. R. Armstrong et al., *Rheolo. Acta,* 20, 163, 1981.
24. H. Laun, *J. Rheology,* 30, 459–501, 1986.
25. W. Gleissele, in *Rheology,* vol. 2, G. Astarita et al., Eds., Plenum Press, New York, 1980, 457.
26. G. Vinogradov et al., *Rheology of Polymers,* Spring-Verlag, New York, 1980, 338.
27. L. Sperling, *Introduction to Physical Polymer Science,* 2nd ed., Wiley, New York, 1994, 458–502.

28. J. Enns and J. Gillham, in *Polymer Characterization: Spectroscopic, Chromatographic, and Physical Instrumental Methods,* C. Craver, Ed., ACS, Washington, D.C., 1983, 27.
29. J. D. Ferry, *Viscoelastic Properties of Polymers,* 3rd ed., Wiley, New York, 1980.
30. A. Ya, Goldman, *Prediction of the Deformation Properties of Polymeric and Composite Materials,* ACS, Washington, D.C., 1994.
31. M. L. Williams, R. F. Landel, and J. D. Ferry, *J. Amer. Chem. Soc.,* 77, 3701, 1955.
32. J. Kubat and M. Rigdahl, in *Failure of Plastics,* W. Brostow and R. Corneliussen, Eds., Hanser, New York, 1986, Ch. 4.
33. R. Tanner, *Engineering Rheology,* Oxford Science Publishers, Oxford, 1985, 352–353.
34. N. D'Sousa, in preparation.
35. W. Brostow in *Failure of Plastics,* W. Brostow and R. Corneliussen, Eds., Hanser, New York, 1986, Ch. 10.
36. D. Plazek, *J. Rheology,* 40(6), 987, 1996.
37. M. Baumgarertel and H. Winter, *Rheologia Acta,* 28, 511, 1989. N. Orbey and J. Dealy, *J. Rheology,* 35, 1035, 1991.
38. J. Dealy and K. Wissbrum, *Melt Rheology and Its Role in Polymer Processing,* Van Nostrand Reinhold, Toronto, 1990, 86–100.
39. A good introduction to the manipulation of DMA data can be found in H. Hopfe and C. Hwang, *Computer Aided Analysis of Stress-Strain Response of High Polymers,* Technomics, Lancaster, 1993. This book gives a fuller development of what is discussed here.
40. J. Ferry and K. Ninomiya, *J. Colloid Sci.,* 14, 36, 1959.
41. My favorites are the packages from Rheometric Sciences and the Isis program from Prof. Winters at U. Massachusetts Amherst. Simpler packages can be found in Hopfe (ref. 39 above) and in G. Gordon and M. Shaw, *Computer Programs for Rheologists,* Hanser, New York, 1994.
42. R. Bird, R. Armstrong, and O. Hassager, *Dynamics of Polymeric Liquids,* V1, Wiley, New York, 1987, 143–150. R. Nunes et al., *Polymer Eng. Sci.,* 22, 205, 1982. G. Pearson et al., *Polymer Eng. Sci.,* 18, 583, 1978.
43. R. R. Rahalkar, *Rheologia Acta,* 28, 166, 1989. R. Rahalkar and H. Tang, *Rubber Chemistry and Technology,* 61(5), 812, 1988.
44. See the Proceedings of the Annual Technical Conference of The Society of Plastic Engineers #53, 54, and 55 for example.
45. J. Bonilla-Rios, Ph.D. Thesis, Texas A&M University, College Station, TX, 1996.
46. S. Wu, *Polym. Eng. Sci.,* 25, 122, 1985. A. Letton and W. Tuminello, *ANTEC Proc.,* 45, 997, 1987. W. Tuminello and N. Cudre-Mauroux, *Polym. Eng. Sci.,* 31, 1496, 1991.
47. Rheometrics Science. Discussion taken from information supplied by K. L. Lavanga.
48. Y. Liu and M. Shaw, *J. Rheology,* 42(2), 267, 1998.

8 DMA Applications to Real Problems: Guidelines

This chapter was written at the request of many of the students in my DMA course asking for a step-by-step approach to deciding which type of test, what fixtures, and what conditions to use. The following was developed to formalize the process we go through in deciding what tests to run. The process is shown in flow charts in Figures 8.1, 8.2, and 8.3.

8.1 THE PROBLEM: MATERIAL CHARACTERIZATION OR PERFORMANCE

The first question that has to be asked, and often isn't, is, "What are we trying to do?" There are two basic options: one could characterize the material in terms of its behavior or one can attempt to study the performance of the material under conditions as close to real as possible. Several things need to be considered. First, what do I need? Do I need to understand the material or to see what it behaves like under a special set of conditions? If I am interested in performance, is it even possible to test or model those conditions? Sometimes it isn't. Airbags open at about 10,000 Hz and no mechanical instrument generates that high a frequency. To reach that frequency, we would need to use DEA or superposition data. If we decide to do performance tests, we have another choice.

8.2 PERFORMANCE TESTS: TO MODEL OR TO COPY

The reason for running a performance test is to collect data under conditions that duplicate or approximate use. This is often done when one knows what material works, but is unsure of what material parameters make a good material good. In many mature industries, a material is optimized by trial-and-error over many years, and no laboratory test is used to study it. Sometimes, the end-use conditions are so complex or require the interaction of so many variables, you don't trust a characterization approach to give you useful information. And sometimes you just want to see how something will work.

One approach is to carefully measure the stresses, the frequency, the temperature range and changes, and even the wave shape of the process. One then applies as exact a copy as one can and records how the material responds. This approach has some problems, as the time scale or the frequency may be outside of the instrument's limit. It also requires a high degree of understanding of the process so that you can copy the key step.

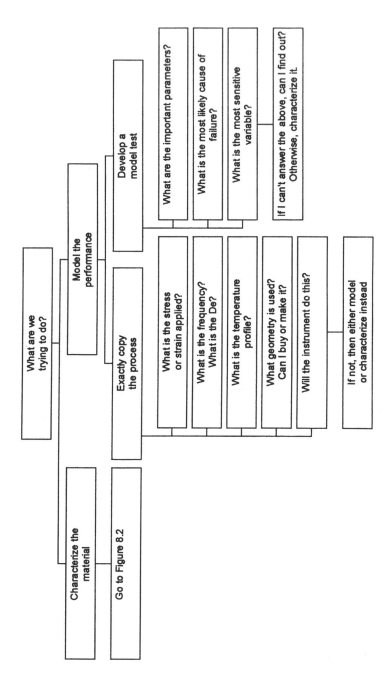

FIGURE 8.1 Choosing a DMA test method requires deciding whether we want to characterize the material, copy the process, or evaluate a model for significant properties.

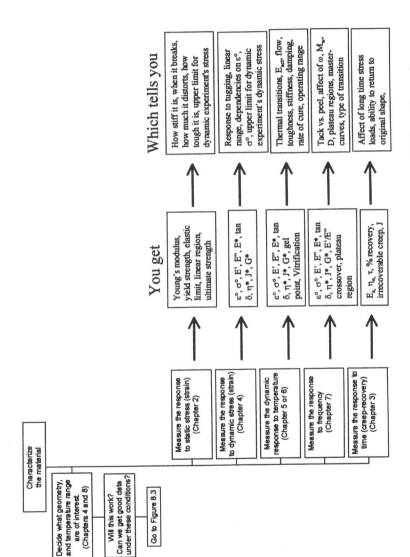

FIGURE 8.2 **Characterizing a material** allows us to gain a tremendous amount of information on it.

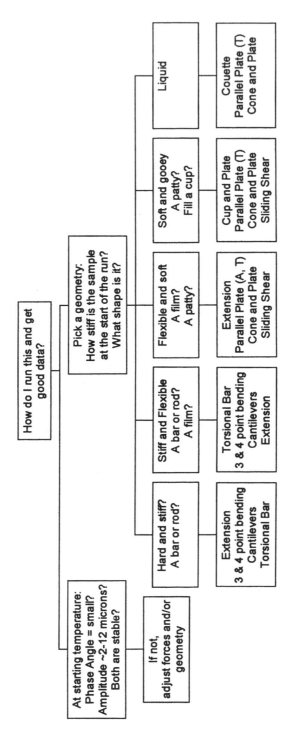

FIGURE 8.3 Selecting the proper test geometry.

TABLE 8.1
Test Methods used in the DMA

Test	Result	What it tells you
Thermomechanical Analysis	Changes in slope	Transition temperature
	Slope of curve with temperature	Thermal expansitivity or CTE
	Change in volume (dilatometry)	Shrinkage on curing, volumetric expnsion
Static Stress-Strain	Slope	Young's Modulus
	Yield Point,	Strength before distortion
	Yield Strength	Load capacity
	Proportional limit	End of linear region (max. F_T)
	Ultimate Strength	Strength at breaking point
	Elongation at break	Ductility
	Area under curve	Toughness
Dynamic Stress-Strain	Dynamic Proportional limit	End of linear region (max. F_D)
	Storage, Loss Modulus	Stiffness as function of load
	Complex viscosity	Flow under dynamic load
	Tan δ	Damping
	Ultimate strength	Strength at break by tugging
Creep-Recovery	Equilibrium Compliance, Modulus, Equilibrium	Long term behavior
	Viscosity	Extensional Viscosity
	Creep Compliance	Effect of load
	Creep Rupture	Strength
	Relaxation spectra	Molecular Modeling
Stress Relaxation	Compliance and Modulus	Long term behavior, MW, crosslinks, and entanglements.
	Retardation spectra	Molecular modeling
	Force as a function of temperature	Shrinkage/expansion force
Dynamic Temperature/time Scan	Storage and Loss Modulus	Change in stiffness. Mapping modulus
	Complex Viscosity	Change in flow
	Tanδ	Damping, energy dissipation
	Temperature of transitions as drops or peaks	$T_M T_g T_\alpha T_\beta T_\gamma T_\delta$
	Modulus of rubbery plateau	Molecular weight between crosslinks, entanglements
	Crossover of E' and E" on curing	Gel point
	Shape of viscosity curve on curing	Miminum viscosity, E_{act} Vitrification point
Dynamic Frequency Scan	Complex viscosity, loss modulus	Flow as function of frequency
	Storage modulus	Elasticity or stiffness as function of frequency
	E'/E" or η* crossover	Relative MW and D
	Plateau regions	MW estimation
	Mastercurve	MWD, long tern behavior, wide range behavior, molecular modeling

In order to avoid the difficulty of matching the process, many people or industries pick a performance test that either roughly models real life or should give similar results. Some of these are excellent, backed by years of experience and knowledge, while others are poor models that are used because they always have been. For this approach, standard sets of highly controlled conditions are chosen to test representative performance. Table 8.1 gives a list of the tests performed in most DMAs and how they can be used to represent certain processes.

8.3   CHOOSING A TYPE OF TEST

What type of test do we choose? The test needs to reflect the type of stress the material will see, the frequency at which this is applied, the level of stress or strain, and the sample environment. First, it is helpful to know what the stress, strain, and strain rate are for the process. This will be needed to see how the material acts. How fast are the stresses applied to the material, are the stresses steady (or constant), increasing, decreasing, or oscillatory? If the stresses are oscillatory, are they sinusoidal, step-wave, or something else? How large are they and how fast are they applied? Once the type of stresses or strains are defined, we can choose from the tests listed in Table 8.1.

The temperature is the next question. This needs to include any heating or cooling the sample would see. Even something like baking a cake has a temperature ramp as the material comes to equilibrium in the oven. Often the material will see heating and cooling cycles that may change the forces applied to it. (See compression set discussion in Chapter 6). The thermal cycle we choose should copy these changes. Sometimes we find that the time at a set temperature is less important than the getting to and from it. Other times, it is the exposure of the material to a specific gas at a elevated temperature that causes the problem, and we may need to use gas switching.

The shape of the stress wave is important and can vary considerably. For example, brushing hair (and modeling the degradation of hair spray) can be seen as a square wave, a heart beat is a composite wave, and vibration can often be a sine wave. The sinusoidal wave works for a lot of processes, but for very nonlinear materials, like those used in some biomaterials, it helps to match the wave shape as closely as possible. People working with heart valves and pacemakers often desire complex waveforms that closely match those of the heart itself.

8.4   CHARACTERIZATION

The alternative to performance tests is to characterize the material as fully as possible. This has the advantage of picking up differences that are not readily seen in the test designed to measure performance. These differences might affect a material during long-term use or change how fast it degrades. In addition, since we will try to relate the tests to various molecular properties and processing conditions, this approach allows us to understand why materials are different and how to tune them.

8.5   CHOOSING THE FIXTURE

The type of fixture you pick is driven by the modulus of the material, the form it is in, and the type of stress the material experiences in use. Certain forms, like fibers and films, suggest certain fixtures, like extension in this case. This is not always true. Sometimes a very thin sample can be handled in extension, three-point bending, or dual cantilever. The choice can then be made on several grounds. I prefer to pick the fixture that is the easiest to load and gives the most reproducible data. Another argument will be to use the fixture than gives the cleanest type of deformation. Yet another is to use one that is similar to the real-life stress of the material. This sometimes requires making a specially shaped geometry, like a pair of concentric spheres to model the eye for contact lenses. Any one of these techniques will work, depending on how the problem is approached.

Table 8.2 lists the common fixtures and the type of samples for which they are most commonly used. Normally, just flexing the sample between your fingers will

TABLE 8.2
Samples and Fixtures

Very Hard	3-pt. bending (large)	Gooey	Various Parallel Plates
	4-pt. bending (large)		Cone and Plates
			Torsion braid
Hard	3-pt. bending (large)	Fluid	Cone and Plate
	Torsion bar		Parallel Plates
			Couette
Stiff and Flexible	3-pt bending (medium)	Film	Extension
	Torsion bar		
Pliable	3-pt bending (medium)	Fiber	Extension
	Dual cantilever		
	Torsion bar		
Soft	3-pt bending (small)	Suspension	Parallel Plate
	Dual cantilever		Cone and Plate
	Torsion bar		Cup and Plate
			Couette
Very soft	Dual cantilever	Powder	Cup and Plate
	Axial Parallel Plates		
	Torsion bar		

lead you to decide the appropriate fixture and the approximate force range. However, you can always check this by loading the sample in the DMA and applying an arbitrary load for 30 seconds. Depending on the modulus value you see, you can then select the most desirable fixture. Figure 8.4 shows the relationship of the various test geometries to modulus for an axial instrument. Some of this is intuitive: stiff hard samples go into three-point bending, rubbery samples into parallel plates, and fluid ones into cups.

8.6 CHECKING THE RESPONSE TO LOADS

Besides being a more accurate method of determining how hard a material is for picking test conditions, the response of a material to a load measured as either stress or strain is one of the most basic studies. For solids, a stress–strain curve lets us see the region of linear behavior, the amount of force needed to make the sample yield, and the force needed to break it. This test allows one to see how a sample responds to a load. However, samples that will be held under loads need to be tested by creep, as there will usually be time-dependence in the response.

8.7 CHECKING THE RESPONSE TO FREQUENCY

As discussed at length in Chapter 7, all materials exhibit some sort of response to the frequency of testing. The frequency of testing affects the temperatures of transitions as well as the modulus and viscosity of the material. Three approaches are normally used to pick the frequency of the test: (1) use an arbitrary number for all

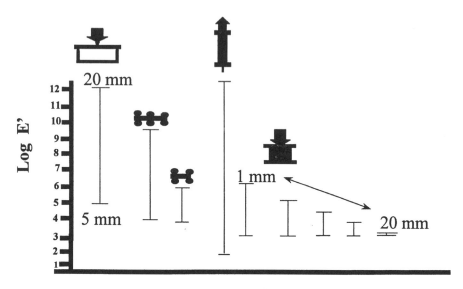

FIGURE 8.4 Modulus range of axial fixtures. A wide range of moduli can be studied simply by varying the fixtures. (Used with the permission of the Perkin-Elmer Corp., Norwalk, CT.)

tests (1 Hz and 10 rad/s are common), (2) measure your processes shear rate and pick the corresponding rate for testing, and (3) run a frequency to pick a testing frequency. One can ideally pick a frequency on the zero-shear plateau, but often these are too low to be useful. The frequency normally ends up in the power law region due to instrument limits.

The third approach is also a good idea even if you have already decided to use one of the first two. How quickly the viscosity and elastic modulus change with frequency is very important for knowing how the material will respond. One needs to remember that frequency and temperature effects overlap and that frequency effects need to be studied at the temperature or in the temperature range of interest.

8.8   CHECKING THE RESPONSE TO TIME

Under this heading, we want to make sure that the material is first stable under the test conditions. This is done by loading the sample at the chosen forces, frequency, and temperature. We should be seeing a level of deformation and strain percentage of under 0.5%. The sample is then held for 1–5 minutes under these conditions and tan δ, E', and E'' are examined. If an upward or downward trend is seen, it means the material is changing as a function of the test conditions. This can be due to extremely long relaxation times or to being out of the elastic limits. One approach would be to remove the testing forces and let the material relax after mounting to see if that helps. Creep–recovery testing is also done to examine the relaxation response.

Another point of this test is that any mechanical or environmental noise in the vicinity will be seen in the test. This allows us to find and remove sources of error that have nothing to do with the test. A lot of the oddities in data are caused by environmental effects including mechanical vibration, impure gases, poorly controlled gas rates, improper cooling, or noisy power. All of these sources of error must be eliminated before a test can be considered valid.

As mentioned above, creep–recovery testing is also done to see how time affects the polymer. This is done to determine the linear region for creep–recovery curves and to measure relaxation times. One first applies creep stresses to the sample in increasing amounts and plots this as compliance, J, versus time. In the linear region, J becomes independent of the stress, so the curves overlay. A fast way to check this is to start at a very low stress and increase it by doubling the stress for each run. When the strain stops doubling, we are out of the linear region.

8.9 CHECKING THE TEMPERATURE RESPONSE

Using the information from above, we now check the response of the material to temperature by running a temperature scan. One normally scans the widest range possible within instrumental limits. The end of the linear region from the static stress–strain curve tells us the maximum total force and the maximum from the dynamic stress–strain run tells us the maximum dynamic force. Then, we pick a dynamic force strong enough to give us a strain between 2–12 μm. If we are running a solid sample, we adjust the static or clamping force to keep the specimen in good contact with the probe. For a hard glassy sample, this is normally 110% of the dynamic force. As the sample becomes soft and more rubbery, this increases as necessary. For fluid samples in a torsional rheometer, we set the gap to where the sample is able to keep a smooth edge. Obviously, for some geometries such as the cone-and-plate, the fixture is designed to run at a fixed gap.

We adjust the positioning of the solid sample to get as low a phase angle as possible at those stresses. (Remember, we are concerned about the stress, not the force.) This is especially important in cantilever and extension geometries, where it is easy to misalign the specimen. The specimen needs to be set up at the lowest temperature, as otherwise the forces may not be sufficient. The sample is then run at a heating rate slow enough to give even heating to the specimen. This should allow one to identify the transitions in the material. It is not uncommon to need multiple runs or special control to collect all these data, as some materials will change so much at the T_g that the specimen will fail.

8.10 PUTTING IT TOGETHER

Now that we have collected the data, we need to apply the analysis given in the previous chapters to determining what it means for our sample. We should now be able to determine the linear region, the effects of temperature, stress, and frequency, and how time-dependent the material is. In addition, we should be aware of where transitions occur and to what of type of behavior they correspond.

8.11   VERIFYING THE RESULTS

So how do we know that the data are good? The first thing that needs to be done is to verify the calibration. The calibration files need to be checked and the values verified. On the Perkin-Elmer DMA, running a series of verification tests as discussed in Chapter 4 can do this. Similar tests can be done on other instruments. After being sure the instrument is operating, the tests should be run in triplicate to be sure that the numbers are real. We can then use the data to check for material variations and operator errors.

The data then need to be examined for inconsistencies and abnormalities. I normally look at five printouts when doing this:

1. The method used.
2. A plot of storage modulus, loss modulus, and tan δ vs. temperature.
3. A plot of program temperature and actual temperature vs. time.
4. A plot of amplitude and phase angle vs. temperature.
5. A plot of probe position, static stress, and dynamic stress vs. temperature.

The method used and plot of E' and tan δ are examined to look at transitions and material behavior. The method is checked to make sure that what was supposed to be run was actually programmed. Were the stresses or strains correct? Was the right temperature range chosen? Were the forces turned on? Were any needed special controls actually applied? The data is checked to see if the transitions and behavior of the material make sense. Are there any sharp drops or abrupt changes that suggest an electronic, mechanical, or sample-related problem? Do the E' and E'' differ from those recorded from previous samples? If so, is the tan δ different too? If E' and E'' have changed but tan δ hasn't, check the stresses and strains. Do they match up correctly?

The temperature plots are used to confirm that the analyzer did what was asked of it and that it stayed in control. Sometimes, due to sample mass or heating rate, the sample temperature will not track the programmed temperature and the data are then suspect. Are there loops in the temperature data suggesting furnace instabilities, variations in gas flow, or ice dropping on the thermocouple, which happens occasionally in subambient runs under high humidity?

The plots of probe position, forces, and raw signals (amplitude and phase angle) are examined for abnormalities. Does the probe position show any sharp changes that suggest sample slippage or movement? Did the stresses behave as expected? Do the raw signals appear to match the data? Any discrepancies need to be explained.

It is also advantageous to examine the sample after the run. Does it show marks from excessive clamping force? Is there evidence of distortion or deformation? Does it weigh the same as it did originally? Has its appearance changed? Can these changes have generated spurious transitions in the spectra?

Finally, we need to ask whether this response a real event, environmental noise, or an abnormality. Does the material do this consistently? Is the material variation so great that the sampling will determine the results? Unfortunately, many rheological and thermal tests seem to be done only once; at least three samples should be run to confirm that the data are correct. In addition, the normal statistics of sampling and data analysis need to be followed.

8.12 SUPPORTING DATA FROM OTHER METHODS

DMA does not always represent the best way to analyze a sample. Some changes that influence the DMA spectrum are more easily studied by other methods. For example, changes in degree of crystallinity can be measured in the DSC in one run. Activation energy, degree of cure, and kinetics can also be studied in the DSC. Vitrification shows up in the storage heat capacity curve from the DDSC. Losses of water, decomposition, filler levels, and the presence of antioxidants can also be seen in the TGA or in one of its hyphenated variants. IR is also an excellent way to investigate questions about composition. DEA allows detection of curing past the DMA's instrument limits (both upper and lower) for thermosetting systems. Information from alternative methods is a vital part of the analysis of a material, and it is foolhardy to limit one's approach to DMA (regardless of how much fun it is to use!) or any other instrument.

APPENDIX 8.1 SAMPLE EXPERIMENTS FOR THE DMA

Before running any experiments on a DMA, it is important to check that it is properly set up and calibrated. For example, the purge gas type, the purge gas rate, the cooling system, the last date of calibration, and the last verification tests should be checked. After assuring the instrument is ready to run, the following series of experiments will help you gain experience on how your instrument operates.

TMA EXPERIMENTS

Ideally, the fixture used should be quartz for low thermal expansion. A baseline should also be run on the empty fixtures. Sometimes a quartz microscope slide will be run as a blank and the sample then run on top of it. A sample of polystyrene[1] should be run under a maximum of 10 mN load in nitrogen purge from 25–125°C at 10°C as follows:

1. A solid piece run under a 3-mm probe in expansion to determine the T_g and the CTE.
2. A small bar run with a knife edge probe in three-point bending to determine the T_g.
3. A small slab run with a 0.5-mm probe in penetration to determine the T_g.

After these three runs, compare the effect of the method on the T_g and the range of the region of the transition.

STATIC STRESS–STRAIN SCANS

For a small bar of nylon 66 and another of Polyphenylene Sulfide (PPS), set up in three-point bending or torsion bar, a stress or strain sweep is applied from 0–2000 mN of static load at 100 mN/min ramp rate. For one specimen, repeat this experiment three times on the same sample. Also run a sample under two additional ramp rates (i.e., 200 and 400 mN/min). Finally, one can try the same experiment using single and dual cantilever if time permits. These should all be done at room temperature. Display the data so you can see the effect of ramp rate and multiple ramps on a sample. Calculate the slope of the stress–strain plot to obtain the modulus.

CREEP–RECOVERY EXPERIMENTS

Using a rubber Gas Chromatograph (GC) septum or disk of butyl rubber, set up a simple creep experiment at 40°C under nitrogen purge. Using a small even recovery force such that the material shows no creep, run an experiment where 200 mN is applied for five 1-minute cycles with 3 minutes of recovery time in between cycles. Compare the first and the last cycles noting differences in percentage of recovery. Repeat with 1000 mN creep load and again at 80°C.

CONSTANT GAUGE LENGTH EXPERIMENT

Using the necessary controls, load a single polypropylene (PP) fiber (common fishing line) in extension with nitrogen purge and low temperature cooling and hold it under enough tension to keep it taut. Set the position control to maintain the current position and turn on your static control. Cool this specimen to –50°C and run a 10°C temperature ramp up to 200°C and recool to –50°C. Plot static force, probe position, and temperature as functions of time.

DYNAMIC STRESS–STRAIN SCANS

Using the same materials and same-size specimens as in the static stress–strain experiment above, run a dynamic stress scan using tension control set to 110% and a dynamic ramp rate of 95 mN/min. Compare the results with the static scan at 200 mN/min. Then plot the storage modulus, loss modulus, and tan δ as a function of dynamic strain.

DYNAMIC TEMPERATURE SCANS

Using samples of polystyrene in three-point bending, PET film (from a transparency or coke bottle) in extension, and nylon 6 in a cantilever geometry, run temperature scans from 25–250°C at 10°C/min. Pick a dynamic force that gives a 10-μm starting amplitude or a 0.05% initial strain and set the static force in 110% ratio to it. Calculate the T_g from the tan δ peak and onset as well as the onset of the E' drop. Compare the results for polystyrene with the results from the TMA experiment above.

   If possible, repeat for one sample in strain control setting the dynamic force control to 10 μm of amplitude and with tension control of 110%.

CURING STUDIES

Using a cup-and-plate or a parallel plate geometry and a sample of commercial two-part high-strength epoxy, set up a run to cure the material in the DMA. Assuming the material takes 2 hours to set (the back of the label gives the setting time), run isothermally at 40°C, 50°C, and 60°C. Depending on the viscosity of the initial material, you may want to float the probe and use amplitude control. Find the gelation and vitrification points. Alternately, you can run a temperature ramp at 10°C/min up to 300°C. In either case, you can use Roller's method (as discussed in Chapter 6, Section 6) to estimate the activation energy, which you could also compare to the DSC value.

FREQUENCY SCANS

In either the extension or three-point bending geometry, set up a sample of PVC as described above at RT. Then set the analyzer to vary the frequency across the full range. Repeat this experiment at 50°C and 75°C. Display the data as log E' and log η^* against log ω. The three runs can also be used for a time–temperature superposition if desired.

NOTES

1. All samples can be obtained in convenient form from the Society of Plastic Engineers in their "ResinKit" set. This is a collection of 50 polymer samples used to teach the identification of plastics and is listed in their book catalog.

Index